SCIENCE FAIR SEASON

SCIENCE FAIR
SEASON

Twelve Kids, a Robot Named Scorch . . .
and What It Takes to Win

———————————•———————————

JUDY DUTTON

NEW YORK

Library of Congress Cataloging-in-Publication Data has been applied for.

ISBN 978-1-4013-2379-0

Hyperion books are available for special promotions and premiums.
For details contact the HarperCollins Special Markets Department in the
New York office at 212-207-7528, fax 212-207-7222, or email
spsales@harpercollins.com.

Book design by Fearn de Vicq

FIRST EDITION

10 9 8 7 6 5 4 3 2 1

THIS LABEL APPLIES TO TEXT STOCK

We try to produce the most beautiful books possible, and we are also
extremely concerned about the impact of our manufacturing process on
the forests of the world and the environment as a whole. Accordingly,
we've made sure that all of the paper we use has been certified as
coming from forests that are managed, to ensure the protection
of the people and wildlife dependent upon them.

To my mom and dad,

who taught me to always try my best

CONTENTS

CONTENTS

SCIENCE FAIR SEASON

INTRODUCTION

One thing I have learned in a long life: that all our science,
measured against reality, is primitive and childlike—
and yet it is the most precious thing we have.
—MARK TWAIN

The cockroaches were the size of cell phones. They arrived, alive, via priority mail. After dumping them into a tank, fifteen-year-old Tristan Williams painted numbers on their backs, 1 through 20, to tell them apart. He fed them chicken feed and table scraps, noting that the only food they refused to eat were Krispy Kreme donuts. This made him wonder: *Do these roaches know something about Krispy Kreme that we don't?* Interesting question. But right now, Tristan had more pressing priorities.

It was March, a month that kicked off a series of competitions that Tristan adored more than any other: science fair season. Tristan, a brainy, bespectacled high school sophomore plagued with poor athletic skills, had long ago embraced science as his sport. He'd competed in science fairs since kindergarten and had come in first every year but fifth grade. The secret to his winning formula? Insects, of course. In first grade, having discovered that ants detest the smell of catnip, he created an insect repellant out of the plant that kept flies off his sister's horse. In sixth grade, after reading that pill bugs devour poisonous chemicals such as lead and cadmium, he conducted an experiment proving the bugs could be used to clean up toxic waste sites. In ninth grade, after hearing his mother gripe about a flour beetle infestation in her pantry, he found a way to keep the beetles at bay using buffalo gourds. Tristan's ingenuity in this realm was unrivaled. Trophies and blue ribbons lined his bedroom walls. He'd even ascended the ranks to win regional and state fairs. This year, though, Tristan was determined to take his project all the way to the top.

That's when cockroaches—specifically, Madagascar cockroaches—came to mind. Tristan had read that these insects had two unique things going for them: a keen sense of smell and the ability to hiss loudly when threatened. Since Tristan's hometown of Las Cruces, New Mexico, was located fifty miles from the Mexican border, cops regularly conducted drug checks along highways and in schools. Drug-sniffing dogs, Tristan learned, were expensive, costing $25,000 each to train. This got Tristan thinking: Could Madagascar cockroaches be trained to hiss when exposed to certain odors? And if so, couldn't law enforcement ditch drug-sniffing dogs for the much cheaper alternative of drug-sniffing cockroaches?

It was a flash of science fair genius, the kind of moment Tristan prayed for. Still, a brilliant idea was just the first step. If Tristan's project was to take off, he'd have to tackle a whole series of tricky problems. First up: the drugs. Tristan knew his parents wouldn't approve of him having a stash of marijuana, even for purely experimental purposes. And besides, the use of illegal substances was forbidden at science fairs unless you filled out tons of paperwork. So Tristan cast about for an alternate odor that would still prove his point. Spotting a felt-tipped pen on his desk, he picked it up, uncapped it, and sniffed. *This will work,* he decided.

After meticulously laying out the parameters of his experiment, Tristan was ready to wrestle with an even bigger question: Were roaches trainable? It was time to find out.

Tristan reached into his tank, fished out Roach #1, and placed it in a shallow Tupperware bowl to keep it from scuttling away. Uncapping the pen, Tristan waved it in the roach's vicinity; then he poked the roach with a finger, prompting it to hiss. Tristan repeated this exercise—the pen, the poke, over and over—to see if the roach could be conditioned to hiss in response to the pen's fumes alone. After eighty-five rounds, the roach got the drift, proving that Tristan's hypothesis was right on the money. Elated, Tristan trained twenty roaches total, recorded his results, and wrote a paper describing the potential applications for drug detection. He knew the odds were slim that cops would take note and start replacing dogs with bugs. Still, stranger things had caught on.

Weeks later, at a local science fair, Tristan presented his project. While it raised a few eyebrows, the judges were nonetheless impressed. Tristan was awarded first place, which propelled him to the state level. Tristan won there as well, which meant that finally, with a decade's worth of projects behind him, Tristan had broken into the big leagues. *The New York Times,* catching wind of his experiment, nicknamed him Cockroach Boy. TV news crews flooded his home to film the roaches in action. Tristan even heard that scientists in nearby Los Alamos had played off the same principle and trained a hive of bomb-sniffing bees. There remained just one final hurdle he would need to surmount to claim everlasting fame on the science fair circuit: He'd need to fly to Cleveland, Ohio, and compete at the Intel International Science and Engineering Fair (ISEF).

Intel ISEF is the Super Bowl of science fairs. Every year, more than fifteen hundred high schoolers from more than fifty countries convene to fight for more than $4 million in prizes and scholarships. Given that this was the top of the science fair pyramid, Tristan knew that the competition would be fierce. But only once he entered the convention hall and trolled the aisles did it dawn on him what he was up against. In one corner of the hall, a girl with glasses explained how she'd synthesized a drug that slowed the growth of cancer. In another booth, a team of three boys from Brazil wowed audiences with a *Terminator*-style bionic arm. In another corner, genetically engineered plants promised to put an end to world hunger. And those were just the projects Tristan could figure out. For some, the title alone—e.g., "Endogenous Estrogen Amplification Through Interaction with P450 Enzymes: Novel Mechanisms for Xenoestrogenicity"—was enough to make his head spin.

Taking it all in, Tristan started to sweat. Trained cockroaches would look downright pathetic next to these scientific opuses. In spite of all his hard work, in spite of his patience putting up with the wisecracks and roach jokes from kids in school, Tristan would return home empty-handed. He didn't stand a chance.

Or did he?

Science fairs bring back memories for just about all of us. The petri dishes. The potato clocks. The classic, crowd-pleasing baking soda volcanoes. Back in fifth grade, my science fair project consisted of rocks dug up in my backyard and pasted on a cardboard display. It didn't win, but I was proud of my work. One day, I dreamed of being a scientist myself.

Flash forward twenty-some years. I was a journalist in New York, surfing the Web, when I stumbled across a *New York Times* article about Tristan Williams and his drug-sniffing cockroaches. Something about his story—his creativity, his drive—fired me up. Like a high-IQ *American Idol,* science fairs were a stage where kids like Tristan could strut his stuff. I was rooting for him, and sad to discover, with further digging, that Intel ISEF did not find him worthy of an award. How could this be? Was the competition really that stiff?

On May 10, 2009, I flew to Reno, Nevada, where Intel ISEF 2009 was taking place. As I strolled through the aisles, it didn't take long for me to see why trained cockroaches didn't make the cut. The first thing I spotted that threw me off guard was a nuclear fusion reactor, which fused atoms together using the same principle that powers the sun. A few booths down, I marveled at a student who had genetically engineered "smart worms" that could do things a worm had no business doing. Next to that, one student's nanotechnology project had led to five patents and a company slated to rake in $12 million. At every booth, I had to pinch myself as a reminder that kids—kids!—had come up with this stuff. And I wasn't the only adult who'd roamed through this surreal sea of tri-panels and left impressed.

"High schoolers are now solving problems that have puzzled scientists for years," says Larry Bellipanni, professor of biological sciences at the University of Southern Mississippi, who's attended fairs for more than forty years and conducted studies on their participants. Andrew Yeager, who's judged science fairs for thirty years and is a professor of medicine at the University of Arizona College of Medicine, seconded this notion, adding, "The level of sophistication in these projects is in many cases beyond the level of graduate school and doctoral research." Hoping to tap into this font of creativity, representatives from the National Institutes of Health, federal agencies, elite universities, and other organizations have walked the aisles of science fairs searching for fresh talent and ideas, and they've found plenty.

Take plane crashes. In 2009, after a U.S. Airways plane landed in the Hudson River due to engine failure caused by birds, aeronautics engineers redoubled their efforts to invent a bird radar system that could alleviate the problem. Katie Stine, a seventeen-year-old in Hilton Head, South Carolina, created a much simpler solution: a meshlike metal cone that can be placed over the front of the plane's engine and would allow air through but deflect obstacles in its path. Given that testing her device on a plane wasn't an option, Stine devised a way to simulate the same conditions using a tennis ball machine, firing a flurry of one-hundred-mile-per-hour balls at the cone to mimic bird collisions. The resulting prototype racked up numerous awards at science fairs and may soon have a patent. Which is not at all unusual: One in five national science fair contestants have patents pending for their projects.

Students who hope their science fair projects may one day impact the world aren't being naive. When NASA launched their Galileo spacecraft toward Jupiter in 1991, they planned a picture-snapping flyby of the asteroid 951 Gaspra, using data collected by seventeen-year-old Claudine Madras from Newton, Massachusetts. Treatments for ailments from autism to Alzheimer's have also benefited from the contributions of science fair contestants. After Gainesville, Florida, seventeen-year-old Kyle Jones discovered that a substance called conjugated linoleic acid kills 90 percent of colon cancer cells within three days, the M. D. Anderson Cancer Center used his research as a springboard for further experiments. Even the Pepsi you get from a vending machine may soon be safer thanks to Taylor Jones, a sixteen-year-old from Alcoa, Tennessee, who patented a germicidal light that can kill microbes lurking on the can as it's dispensed.

Cancer cures and cleaner Pepsi cans aren't all that's at stake. Some experts argue that the very future of our country rests on these kids' shoulders. According to a recent report by the International Center for Education Statistics, American fifteen-year-olds scored significantly lower in science literacy than students in China, Japan, Canada, the Czech Republic, and eighteen other countries. The Hart-Rudman Commission, tasked with finding solutions to national security threats, concluded that the failures of our math and science education "pose a greater threat . . . than any potential conventional war." Even Bill Gates is scared, having addressed a summit of state governors by saying, "When I compare our high schools to what I see when I'm traveling abroad, I'm terrified for our workforce of tomorrow."

Science fairs, unlike high school sports, rarely fill the rafters with raving fans. But the tide is turning. In 2009, President Barack Obama announced that the White House would begin holding an annual science fair in an effort to move the United States "from the middle to the top of the pack in science and math over the next decade." After all, the President pointed out, "If you win the NCAA championship, you come to the White House. Well, if you're a young person and you've produced the best experiment or design, the best hardware or software, you ought to be recognized for that achievement, too. Scientists and engineers ought to stand side by side with athletes and entertainers as role models, and here at the White House, we're going to lead by example. We're going to show young people how cool science can be."

As I wandered slack-jawed through the aisles of Intel ISEF 2009, I wasn't just bowled over by the sophistication levels of the projects.

What hit me even harder were the stories behind them. Many of the kids had been inspired to solve problems in their own lives. One boy, whose family was scraping by in a derelict trailer with no heat or hot water, designed a solar-powered heater out of a 1967 Pontiac radiator, sixty-nine soda cans, and other junk he found around town. Another girl, hoping to help her little cousin cope with autism, developed a treatment program that enabled her cousin to read, write, and interact with others, and proved so successful that it was rolled out in schools across the country. In another town, where local cops had been rattled by a rash of suicides, a science fair project involving "therapy horses" helped keep post-traumatic stress at bay.

In addition to tackling the intricacies of particle physics and thermodynamics, these kids encountered obstacles outside the lab that would make even the most stalwart competitor crumble. One girl, after raising awareness about a cancerous chemical being dumped in the water by the multibillion-dollar company DuPont, became a terrorist suspect under investigation by the FBI. Other kids stumbled across more pleasant surprises, like $420,000 in prize money, a trophy case next to the football team's, and the chance to ask a girlfriend to the prom on *Good Morning America*. As I rode along through their highs and lows, I saw not only how these kids were transforming the world, but how the experience was leaving an indelible mark on them and who they'd one day become. Thanks to science fairs, two boys sitting in a juvenile detention facility learned they were smarter than they thought. A girl with leprosy discovered that even the most devastating disease had an upside if you looked for it. Science fairs gave them confidence, and courage, and hope.

Inspired by the kids I met at Intel ISEF 2009, I stayed. Over five days, I watched them set up their projects, field questions from judges, then take their seats at the awards ceremony to see if their dreams of science fair stardom would come true, or crash and burn. Out of 1,502 registered competitors, I met six who captivated me so much, I dedicated six chapters of this book to documenting their journeys. The other five chapters are devoted to past competitors whose stories are the stuff of legend in science fair circles. Curious whether the rumors I'd heard about these students' exploits were true, I made pilgrimages to their homes and laboratories, and was amazed by what I saw.

Science fairs are full of surprises. But there are no guarantees of a happy ending. Some of the competitors profiled in these pages will win. Others will lose. But win or lose, science fairs change kids. And even though I was just an observer, rooting on my six Intel ISEF 2009 competitors from the sidelines, I'd say that science fairs changed me, too.

In a world brimming with bad news—global warming, nuclear proliferation, America's alarming decline in the realm of science education and otherwise—science fairs are the silver lining. They convinced me that maybe we're not as bad off as I might think. While writing this book, I got to know the most hardworking, humbling, and heartbreaking group of young men and women. They changed everything I thought I knew about kids and what they're capable of, and what we can all do if our hearts are into it, at eight years old or eighty.

THE RADIOACTIVE WHIZ KID

A little learning is a dangerous thing.
—ALEXANDER POPE

The first thing that struck me about Taylor Wilson was that he was tiny. At fourteen, his ninety-five-pound frame swam in his T-shirt and jeans. Blond hair framed his blue eyes in a bowl cut. His voice was high as a choirboy's, made even more adorable by his polite Southern drawl. *Yes, ma'am, this is my first year participating in science fairs. Yes, ma'am, I'm having a wonderful time.*

Taylor was one of the first competitors I met at the Intel International Science and Engineering Fair 2009. I was drawn to him for the same reason I gravitate toward puppies and Pokemon characters: Taylor was cute. Only as I spoke with Taylor about his research—and he started trotting out more ominous words like *uranium, terrorism,* and *fissionable radioisotopes*—did I come to the distinct realization that Taylor probably hated being called cute. Taylor wanted to be taken seriously. He was out to prove that kids could do amazing things if they set their minds to it. At first, I admit, I was skeptical. His parents, Kenneth and Tiffany, were once skeptical, too. That all changed one day four years earlier when Taylor invited them into the backyard to see what he'd built.

As Kenneth and Tiffany obligingly followed their son outside, Taylor proudly presented a plastic pill bottle, stuffed with sugar and some of his dad's stump remover. Stump remover, Taylor had read online, contained potassium nitrate, and potassium nitrate mixed with sugar will explode when lit. Poking out of the top of the pill bottle was a fuse. With a flourish, Taylor struck a match. It all happened so fast that Kenneth and Tiffany didn't know what to think. The pill bottle, after all, was so small. Even if it did explode, there was no way it could do any serious damage, right?

Within seconds, there was a thunderous clap. Even the neighbors heard it, and they came running outside in a panic. As they turned toward the Wilsons' backyard, they saw a miniature mushroom cloud rise toward the sky. It was small as far as mushroom clouds go. But then again, Taylor was ten. And what his parents didn't realize yet was that he was determined to build something bigger.

As far as I could see, Taylor came from a nice, mild, down-to-earth Southern family. Kenneth, a tall man with sandy hair and an easy smile, worked in sales at Coca-Cola. Tiffany, who was willowy and soft-spoken, taught yoga and ran a health food store. They lived in Texarkana, a tiny town that straddles the border between Texas and Arkansas, and had two other children—Ashlee, nineteen, and Joey, eleven.

Before building bombs, Taylor's niche was family entertainer. He sang constantly, even in school, and danced in a style all his own, his arms and legs whirling like the blades of a helicopter. Taylor progressed through the typical stages of what a boy wants to be in life, although the intensity with which he pursued these interests struck his parents as extreme. At age three, he fell in love with construction gear. For Christmas, he asked for a hard hat, fluorescent vest, and orange cones—real ones, Taylor detested toys—and used them to direct traffic on his street (which, thankfully, moved at a snail's pace). At age seven, he decided to become an astronaut. Up in his bedroom went a poster showing every rocket ever made by NASA and the Russians, from the 1930s onward. Within days, Taylor could recite this list by heart.

At age ten, Taylor received a gift from his grandmother that, in hindsight, she probably regrets giving him to this day. *The Radioactive Boy Scout* was a true story about a teenage boy named David Hahn who, in the early 1990s, tried to build a nuclear reactor in a potting shed in his backyard. Given that the boy nearly nuked his entire neighborhood before the Environmental Protection Agency arrived in hazmat suits to dismantle his operation, the book was clearly a cautionary tale: *Don't try this at home, kids.* Still, in Taylor's mind, it was hard to ignore the fact that this book also taught kids that if they *did* try, they *could* build a nuclear reactor. That is, if they were smart enough to figure it out.

Up next to Taylor's poster of NASA rockets went a periodic table of the elements. Within days, Taylor could rattle off their atomic numbers, masses, and melting points like most boys recite memorized baseball statistics. While Taylor was familiar with many of the elements already—hydrogen, helium, calcium, copper—he was particularly taken with the thirty-four at the bottom of the chart, since they all shared one interesting trait: They were highly radioactive. Some rang a bell with Taylor already, like uranium and plutonium. Others were more exotic. Polonium was used to poison Russian spy Alexander Litvenenko. Radium was once imbibed as an aphrodisiac, before its adverse health effects became known and it was pulled from the market. Certain elements higher up on the periodic table, while not being exclusively radioactive, came in radioactive forms called radioisotopes. Hydrogen, for example, had a radioisotope called tritium. Unlike hydrogen, whose nucleus contained one proton, tritium's nucleus contained one proton and two neutrons, which broke down and gave off radiation.

Taylor learned that radiation was all around him, and that while some of it was dangerous, a lot of it was not. He was surrounded, for instance, by radiation emanating from the earth and outer space, known as background radiation. Potassium, an element found in your garden-variety banana, was also radioactive, as were Brazil nuts. Some vintage dishes were painted with a radioactive uranium glaze, although the people who ate off them seldom knew and were no worse for the wear. Many forms of radiation were also helpful. Radiation could cure cancer and x-ray broken bones. Radiation could be used for good or evil. It could save millions of lives or destroy them with the push of a button.

Taylor was fascinated. By what? He wasn't sure exactly. Maybe it was that atoms seemed so small and unassuming, but possessed amazing powers, much like Taylor himself. For whatever reason, Taylor wanted to know more. But first, he would need his dad's help.

"Hey, Dad. Could I get a Geiger counter?"

Texarkana is a small town, and Taylor's dad, Kenneth, is a friendly guy. He knew plenty of his neighbors on a first-name basis, and one of them happened to be the man who ran Texarkana's

Office of Emergency Management. In addition to protecting the town's citizens from storms, floods, and flu outbreaks, the office was also responsible for detecting nuclear threats. As such, Taylor reasoned, the office must have an old Geiger counter lying around, most likely gathering dust. If so, might Taylor be able to borrow it?

This question, like most questions Taylor was asking these days, made Kenneth uneasy. Still, not one to deny his son anything without mulling it over, Kenneth made the call. To his relief, his friend in emergency management said yes, they did have a Geiger counter, and no, unless Taylor was *really* creative, he'd have a hard time hurting himself with it. Hearing this, Kenneth figured there was little harm in indulging his son's request. After all, where in Texarkana would Taylor find anything radioactive enough to make the Geiger counter work?

Days later, a Geiger counter about the size of a lunch box was sitting in Taylor's lap. He and his mom, Tiffany, were driving to Hot Springs, a nearby town known for being the hometown of former President Clinton. Antique shops dotted the tree-lined streets, and while Taylor had never expressed any interest in antiquing before, Tiffany was happy to humor him. As they poked around dusty corners filled with ornate armoires and Vienna clocks, Taylor kept his eyes peeled for orange pottery known as Fiestaware. Once he spotted a stack of plates, Taylor made a beeline toward it.

One thing that few people know about Fiestaware is that the orange glaze painted on it contains uranium. Fiestaware, as a result, is radioactive. No matter how old this pottery gets or how much dust it collects, it emits a steady stream of subatomic particles that a Geiger counter detects with a series of clicks. The faster the clicks, the closer and stronger the radiation source. As

Taylor approached the plates, his Geiger counter started clicking. Once he got within arm's reach, the clicks kicked into high gear, melding into a constant *rrrrrrrrr* that got the shop owner craning his neck toward the back of the store to see what was going on.

"Your pottery's radioactive," Taylor triumphantly informed the shop owner, who appeared confused, and then somewhat relieved when Taylor left the store, plate in hand. Taylor had landed his first radioactive item, but his collecting spree didn't end there. After scaring every antique shop owner in town, he hit hardware stores, having learned that some smoke detectors contained a radioactive element called americium, and camping lanterns another called thorium. On sites like eBay, there was no end to the oddities Taylor could purchase. Radon sniffers, nuclear fuel pellets, lead pigs, and spinthariscopes soon rounded out his collection. Even though Taylor was still "borrowing" a Geiger counter from the Arkansas Office of Emergency Management, he acquired thirty Geiger counters, of varying abilities. One, called the Super Scintillator, was so sensitive that the military used it to detect radiation levels on the ground from a plane. Back when it was manufactured in 1955, it cost $1,995, the same price as a Chevy.

For two years, Taylor's collection grew. But even then, he wasn't happy. He didn't just want to *have* radioactive items; he wanted to experiment with them. And that would mean he'd have to embark on his biggest challenge yet: building a Farnsworth fusor.

Philo Farnsworth was born in 1906 and grew up on a farm in Beaver County, Utah. At age twelve, he built his family a mechanical washing machine. At fourteen, he conceived of the television, and he invented the first working model seven years later. Farnsworth should be famous. But lack of money, combined with a patent dispute with RCA, muddled his name in obscurity. Philo continued inventing, and by his death, he had accumulated 165 patents. One of his last, which barely made a blip on most people's radar, was called the Farnsworth fusor.

True to its name, the Farnsworth fusor fused atoms together. Ever since the discovery of fusion, scientists had hailed its potential as a "clean" energy source, since it produced much smaller amounts of radiation than fission, or the splitting of atoms. The biggest hurdle to achieving this goal was that immense amounts of energy were required to fuse atoms together; then, the energy given off was hard to recover. In short, energy input exceeded output. For fifty years, scientists spent billions of dollars trying to reverse this dynamic. But so far they're still scratching their heads.

In the 1990s, nuclear hobbyists called "fusioneers" began building fusion reactors in their garages and basements. On the Internet site Fusor.net, they swapped parts and advice, much like vintage car enthusiasts might trade maintenance tips and engine components. While a few fusioneers chipped away at unlocking the secret to clean energy, far more built fusors for their by-product: neutrons. Neutrons, a form of radiation given off during fusion, served as the building block to an array of nuclear experiments. Shoot neutrons into topaz—a mineral that, in its natural form, is often brown in color—and you turn it blue. Pelt an oil painting with neutrons, and it can help you determine the chemical makeup of the paint without damaging it, which is useful for identifying fakes. Neutrons also have plenty of commercial uses, like the production of radioisotopes, which can be used in radiation therapy and as medical tracers. Any fusioneer who dreams of dabbling needs a steady supply of neutrons, and building a nuclear fusion reactor is his ticket.

Among those who embarked on this quest, there was a pecking order. The rookies were part of the "Scroungers List," those who were assembling parts. The next step up was the "Plasma Club," which included individuals who'd built a "demo fusor" that could create plasma, considered a stepping-stone to fusion. The highest echelon members could reach was the "Neutron Club," which meant an individual had built a fusor that could successfully fuse atoms together and produce neutrons. In 1999, the Neutron Club had two members. In 2001, three new members were added. By the time Taylor stumbled across this list in 2007, there were thirty people worldwide in the Neutron Club. Taylor, at age twelve, vowed to become the thirty-first.

Taylor started emailing Fusor.net members, asking for spare parts and advice. As was the case with any potentially dangerous hobby, Fusor.net's members were wary of newcomers, and twelve-year-old kids were a particularly worrisome presence. Some argued their site put children at risk. Others argued that giving advice was a safer alternative to withholding it and letting kids fumble around on their own. Taylor's grasp of nuclear physics struck a few members as unique.

Eventually Taylor's queries caught the eye of one Neutron Club member named Carl Willis. Carl, a twenty-seven-year-old grad student living in Albuquerque, New Mexico, looked like the type of buttoned-down, bespectacled young man most would be glad to invite to dinner. And yet, at age twelve, Carl had built his first explosive out of Clorox bleach. During college, he whipped up a batch of gunpowder and tried drying it in the dormitory microwave. Within seconds, a fireball erupted from the machine, prompting a mass evacuation and a flurry of concern about his experiments. The college chemistry department, sensing Carl would continue with or without their help, gave him twenty-four-hour access to their facilities, which seemed safer than letting him tinker around in his dorm room. In the lab, with guidance, Carl's talents took off. At age twenty-two, he built a nuclear fusion reactor and became the tenth inductee into the Neutron Club.

While Carl had been lucky enough to find a few mentors who cultivated his interests, the majority of adults in his life viewed his activities as something deviant that should be discouraged. So when he spotted Taylor's pleas, Carl felt compelled to help. He started emailing Taylor relevant papers. Over time, emails turned into phone calls, and phone calls into joint field trips. The twosome toured Bayo Canyon in New Mexico, where test explosions of radioactive material were conducted in the 1940s. They toured the LANSCE particle accelerator in Los Alamos, where Carl had interned one summer. They compared their collections of radioactive items; Taylor's rare finds often made Carl envious. Carl, though, had one thing that Taylor did not: a nuclear fusion reactor. Carl was a card-carrying member of the Neutron Club, while Taylor was just a lowly Scrounger.

Carl knew that many Scroungers eventually gave up. He didn't want to see that happen to Taylor. And so, as a gesture of his faith in his friend's abilities, Carl approached his boss. Carl worked at a company that manufactured particle accelerators. Sitting in their storage room was a high-voltage power supply—a crucial component to building a fusor and a prohibitively expensive piece of the puzzle, given that new ones cost $5,000 to $10,000. Taylor had searched high and low and had yet to get his hands on one.

"I know someone who could really use that high-voltage supply," Carl said to his boss. His boss hemmed and hawed, but with more wheedling, he handed it over. Upon hearing what Carl had acquired for him, Taylor was thrilled. Now that he had this key component, the question remained: Where would he get the rest of the parts?

The answer, Taylor would be surprised to find, was hiding in Reno, Nevada.

Tucked between the casinos, strip clubs, and bars of Reno, there was a hidden enclave of kids who didn't fit in anywhere else. Max Oswald-Sells, at age three, had learned how to read without being taught the alphabet. Misha Raffiee, at age two, had begun playing the violin, and at twelve she became the youngest musician on contract with the Reno Philharmonic Orchestra. Every day at school, Max and Misha sat alongside chess champions, math whizzes, and spelling wunderkinds. While their talents ran the gamut, the one thing these kids all had in common was that they were brilliant—so brilliant, it was sometimes debilitating. Before coming to Reno, Max had been teased and beaten on the playground. In class, he had been bored to tears. His mother, who'd already allowed Max to skip kindergarten, was at a loss about what to do. So when she heard that a school called the Davidson Academy, in Reno, might be able to help, it didn't matter that she lived in Sydney, Australia. She booked their tickets.

In many ways, the Davidson Academy looked like any other small public school. Lockers lined the hallways. Outside, kids played soccer during recess. And yet the school's founders, Jan and Bob Davidson, had a very unique agenda in mind for their students. The Davidsons, who lived near Reno, had made their fortune in the educational software business. Now, as full-time philanthropists, they wanted to reach out to kids who desperately needed their help: the nation's brightest. In this No Child Left Behind era, they noticed, schools were spending the majority of their time and resources bringing kids at the bottom of the grade curve up to par. Meanwhile, the needs of kids at the top were being all but ignored, and they were tuning out—and dropping out—in droves. While statistics varied, one study suggested that as many as one-fifth of high school dropouts scored in the top 1 percent on achievement tests.

Since 1999, the Davidsons had been helping these kids find support and resources with which to excel. Still, since these families were scattered all over the world, they began clamoring for a school, saying they'd happily move so their kids could attend. In 2006, after finding a suitable location on the University of Nevada Reno campus, the Davidsons opened the Davidson Academy and began calling for applicants. To qualify, kids had to score in the top 99.9th percentile on an SAT, ACT, or IQ test. Classes would be free, since the academy was a public school, the first of its kind in the country. In an effort to encourage students to proceed at their own pace, there were no grade levels. Instead, kids were grouped by ability and interests rather than age. It was a novel idea, a science experiment in its own right, a veritable petri dish of budding Einsteins, Beethovens, and Da Vincis.

Back in Texarkana, Arkansas, the Wilsons were pondering whether they should make the move to Reno. Kenneth and Tiffany had heard about the Davidson Academy after their daughter, Ashlee, stumbled across an article about it in *Time* magazine. "Taylor and Joey might like this," Ashlee told her parents, who figured there was little harm in letting their boys apply. They had a hunch their sons were smart, but doubted their sons were *that* smart. Yet when both Taylor's and Joey's IQ test results came back in the 99.9th percentile, the Wilsons had to face the fact that their kids really were different—and, in Taylor's case, perhaps even dangerous.

The Wilsons had prayed that radioactivity was just a phase. But over the past four years, Taylor's obsession had grown, and it scared them. More worrisome still, Taylor's interest in school had begun to wane. Even though he consistently came home with straight A's, he slept through class. At home, he had amassed a sizable pile of pieces required to build a nuclear fusion reactor. Sooner or later, he would put them together and turn it on. What then?

Rather than sit tight and find out, the Wilsons packed their bags. In Reno, Taylor would meet another man who, like Carl Willis, would take him under his wing.

As far back as the 1980s, if a tech company went belly-up, you could bet that Bill Brinsmead would arrive on the scene to clean up the mess. After receiving an inside tip that so-and-so firm was closing its doors, Bill would rent the largest Ryder rental van he could find, drive it all night, and pull up at the company's loading dock. Then he would hop out, shoot the breeze with the security guard, and meander through the abandoned hallways, scrounging for parts. Bill was a senior engineering technician at the University of Nevada in Reno, and the university desperately needed a new computer lab. Why pay top dollar for new equipment when startup companies were crumbling and willing to unload their outcasts for free?

Bill, a former military man with a brawny build, shaggy hair, and shaggier mustache, had earned the nickname the "Pirate from Nevada" since he tended to plunder the spoils swiftly and leave nothing behind. Occasionally, he arrived so quickly after heads had rolled that he'd spot half-eaten hamburgers sitting on desks next to dried-up cups of coffee. Given that many of the items Bill hoped to take weighed thousands of pounds, he often brought buddies to help him shoulder the load, but even then it was back-breaking work. After he'd removed the goods, Bill would politely sweep the area—an added touch that kept him in guards' good graces and all but guaranteed he'd get another phone call as soon as the next company crashed.

In many ways, Bill was nothing more than a high-end janitor removing extremely expensive trash. But through the years, he had outfitted the University of Nevada with some surprising prize finds. Once, he nabbed a machine from Hewlett-Packard that gold-plated semiconductors. By recruiting some physics students to help him dismantle the carcass, he managed to extract $30,000 of gold from the machine, which he used to outfit his university with a departmental server. Another time, Bill arrived on campus towing an electric bus, which he turned into a mobile science Exploratorium that made the rounds to local schools. Bill's biggest find, which required three railcars and two eighteen-wheelers to tow home, was a pulsed power generator called a Zebra, from the Los Alamos National Laboratory, which was turned into a new research branch on campus called the Nevada Terawatt Facility. Whenever Bill's finds didn't have an immediate use, they went into storage until a purpose could be found for them. When certain administrators complained about the clutter, Bill would step in and guard his turf. Someday, he'd point out, that Conflat fixture could come in handy.

Outside of work, Bill also enjoyed scavenging at government auctions for parts to funnel into his personal projects. Among these projects were electric vehicles, or EVs, including a little red electric sports car that could accelerate up to one hundred miles per hour. Since Bill was a die-hard fan of Burning Man, a week-long annual art festival held in the middle of the northern Nevada desert, he'd built an EV just for the occasion: a ten-foot-long replica of the atom bomb Little Boy, topped with a Western saddle. Every year, Bill rode Little Boy through Burning Man in homage to the movie *Dr. Strangelove,* in which a B-52 pilot saddles up an atom bomb and rides it straight down to Russia. Bill had a healthy sense of humor when it came to nukes. That's why the Davidson Academy was banking that he'd be the perfect match as a mentor for Taylor.

Weeks before this odd duo met face-to-face, Bill had received emails from Taylor saying he would soon be attending the Davidson Academy and that he hoped to build a nuclear fusion reactor with Bill's help. At first glance, Taylor's emails were so steeped in sophisticated terminology, Bill assumed someone must be pulling his leg. *Come on, I wasn't born yesterday,* he thought. *Who's this kid's ghostwriter?* After meeting face-to-face with him, Bill wondered whether Taylor was biting off more than he could chew. Still, Bill was reminded of a time back when he was about Taylor's age, stuck sitting in science class, feeling bored and frustrated. Gifted programs were available for reading and math, but there was no such thing as a gifted program for science. Most of his teachers had no idea how to help him build the things he dreamed of building. In eighth grade, Bill managed to piece together a laser, which he entered in his middle school science fair. He lost to a girl who'd fed her guinea pigs vitamin C, and who had put together a prettier poster. Soon after that, Bill's interest in science fairs ground to a grudging halt.

Now, staring down at Taylor thirty-odd years later, Bill had a choice. He could steer this kid toward a science fair project more appropriate for his age, like a hydrogen car. Or he could see if he and this kid could create something really, really cool.

"You're on, kid," Bill said. "Let's go take a look at my collection and see what we've got."

Bill and Taylor meandered from room to room, scanning shelves, grabbing pieces of scrap metal. "This'll work . . . this'll work . . . ," Bill mumbled occasionally, consulting Taylor when he needed more specifics. They piled their finds in the university's subbasement, which was deemed the safest place to start putting these things together. Around the area where they'd be conducting their experiments, they erected a shield made of paraffin and lead, which would absorb any radiation the experiments might produce. Since lead dividers were expensive, Bill came up with a cheaper makeshift alternative: old gel cell batteries from computer power supplies, stacked one on top of the other like bricks.

Once their workspace was up and running, a radiation safety officer stopped by occasionally to assess whether the precautions they were taking were up to snuff. As an added precaution, Taylor and Bill wore dosimeters, which were badges worn by nuclear power plant workers that measured the amount of radiation they'd been exposed to. Taylor, due to his small size and age, was allowed only about half the radiation exposure levels allowed an adult. Just to stay on the safe side, Bill occasionally stuffed Taylor behind him, turning his own body into a makeshift human shield.

After months of researching and welding, they'd assembled a gleaming mass of metal that looked like a cappuccino maker on steroids. Its cylindrical trunk consisted of a reaction chamber, an airtight environment that would serve as the stage for their experiments. Attached were vacuum pumps to remove unnecessary air molecules, fans to keep the machinery from overheating, and last but not least, a tiny window so they could see what was going on inside. Hovering front and center in this window was a Ping-Pong ball–sized framework of wires known as the grid, which was attached to the high-voltage power supply, compliments of Carl Willis. Once everything was in place, it was time to turn the contraption on. Only then would Taylor know whether he could ascend from the lowly ranks of the Scroungers List to the next tier: the Plasma Club.

Taylor had already had some experience with plasma. It had happened at lunchtime, around a month earlier, in the Davidson Academy cafeteria. Taylor, as usual, was eating healthy, a habit that came as a surprise to his classmates, given his other proclivities. That day, he had special plans for one particular item of food on his plate: a grape. As his fellow classmates munched their sandwiches and looked on with mild curiosity, Taylor sliced the grape in half, stuck it in the microwave, and turned it on. While most onlookers assumed that he would end up with a warm grape, Taylor knew better. Within seconds, a fireball formed above the grape, glowing purple.

Grapes, due to their size, shape, and electrolyte content, are great at absorbing and amplifying microwaves, a form of radiation. If placed in the right spot in a microwave, grapes can produce a fourth state of matter known as plasma. Plasma consists of ionized gas, which means the electrons within it roam free from the nucleus they're typically bound to. Stars and lightning are made of plasma, but Taylor had just demonstrated that plasma could also be created through man-made means. His classmates were impressed. His teachers administered a gentle scolding ("No more grapes in the microwave, Taylor"). Now all he needed to do was create those same special effects inside his fusor.

Creating plasma within a fusor was a similar process to the grape-in-the-microwave trick, only instead of a grape, Taylor would be ionizing argon gas, and at much higher voltage levels. As Bill filled the fusor with argon and flipped the switch, Taylor peered through the window into the reaction chamber. At first, all he could see were the tungsten wires of the grid, glowing red. Then, as they amped up the voltage, Taylor saw a bluish haze coalesce around the grid, hovering in midair, like a ghost. This was plasma. Knowing no one would believe him without proof, Taylor snapped some photos, then posted them on Fusor.net. Congratulations poured in, first from Carl Willis, then others. Taylor was part of the Plasma Club and was beginning to earn its members' respect.

A few days later, after tightening a few screws and fine-tuning a few pieces of equipment, Taylor and Bill turned the fusor on again. Only this time, rather than peering into the reaction chamber window, they scooted behind the lead wall they'd erected. That's because this time, they'd loaded the fusor with a different gas, called deuterium. Deuterium, unlike argon, is a fusable gas. That meant that if all went well, Taylor wouldn't merely be creating a blue ball of plasma. That ball of plasma would go the extra mile. The atoms inside it would fuse together.

Fusion is the process that powers the stars. On the sun, hydrogen atoms fuse together to form helium, releasing energy in the process. Replicating this reaction on earth, though, is no easy task. The main problem, Taylor knew, was that atoms don't want to fuse together. Get two of them close enough, and their positively charged nuclei repel one another, like the positive ends of two magnets. Stars use their sheer mass and high temperatures to squeeze atoms together, but Taylor had neither of these brute forces at his disposal. What he did have was his grid, which had two key things going for it. Due to the high voltage levels running through it, it could strip electrons from the nuclei of atoms and form plasma. The grid could also amass a powerful negative charge, much like one end of a battery. Negative charges attract positive ones, and nuclei stripped of their electrons are positive. As a result, the nuclei within the plasma cloud would shoot inward, toward the grid. Some would hit the grid and be absorbed, but others would shoot right through toward the grid's empty center. Essentially, Taylor was masterminding a miniature nuclei pileup. If the speed of the nuclei were fast enough, some of them would crash and stick. They'd fuse together.

Fusion, surprisingly, is a quiet process. There are no big bangs or explosions. The only sound Taylor and Bill could hear from behind the lead wall was a bit of crackling from the electricity source and a couple of blips from a nearby computer. Only after they'd turned off the power supply and rounded the lead wall would they know whether their efforts had been successful, due to a tiny device they'd placed near the fusor called a bubble dosimeter.

When deuterium atoms fuse, they kick out neutrons as a byproduct. These neutrons, while small, carry energy that, if it collides with a bubble dosimeter, can heat the substance inside it (typically Freon) to the boiling point. The result: bubbles. If Taylor saw bubbles in his bubble dosimeter, that would mean he'd created neutrons, and that would mean he'd caused deuterium atoms to fuse together. And that would mean that after four years of scrounging for parts, worrying his parents, and moving across the country, Taylor could finally say it was all worth it.

Taylor picked up the bubble dosimeter and squinted at it: one bubble. He wasn't sure whether to be elated or angry. Was one measly bubble enough to gain entrée into the Neutron Club? Taylor didn't want to leave any room for doubt. Rather than show the bubble to Bill, Taylor suggested they run the fusor again. This time, when they picked up the dosimeter, they counted five bubbles. Five was enough to convince the harshest critic.

That night, Taylor posted proof of his breakthrough on Fusor .net, where he was promptly proclaimed the thirty-first—and youngest—nuclear hobbyist in the world to build a Farnsworth fusor. Taylor's parents, upon hearing the news, celebrated by taking him out to dinner. His teachers suggested that he enter his fusor in the regional science fair, which was coming up in a few weeks.

At the regional science fair, when Taylor and Bill wheeled the fusor into the convention hall, Bill had to chuckle as hordes of students stopped, stared, and tried to figure out what it was. The project's title—"Subcritical Neutron Multiplication in a 2.5 MeV Neutron Flux"—didn't help much. The words *nuclear reactor* rippled through the crowd, followed by *radiation poisoning, gamma rays,* and more whispering. *Better give that booth a wide berth.* Competitors simultaneously resigned themselves to winning second place at best. Bringing a fusor to a science fair was like bringing a Ferrari to a go-cart race.

At the awards ceremony, Taylor won first place and a spot as a finalist at the Intel International Science and Engineering Fair 2009. As luck would have it, the competition would be held in Reno that year. Bill was relieved. That would mean he could drive Taylor's project to the convention center in his electric van, rather than fly it in on a plane. Who knew what airline security would have thought if they'd laid eyes on a nuclear fusion reactor.

Soon after meeting Taylor, I picked up a copy of *The Radioactive Boy Scout,* the book that had launched his quest four years earlier. Reading this book, I was struck by the fact that its main character, David Hahn, was a lot like Taylor. David was in his teens, and obsessed with radioactivity, and determined to build a nuclear reactor no matter what it took. So far, this description fit Taylor to a tee. But at this point, their paths diverged. David's efforts were shut down by teams in hazmat suits. Taylor's story could have a happy ending.

"Exactly," Taylor agreed with me. "Every nuclear scientist I meet who's read that book always says, 'You're the ending we wanted to see to that book.'"

Taylor, like David, had the potential of turning into a parent's worst nightmare. Only Taylor didn't, and the reason for this boiled down to one crucial difference: support. David was alone, angry, bored, and brilliant. It was only a matter of time before his efforts veered into dangerous territory. Taylor, on the other hand, did not build his nuclear fusion reactor alone. A quirky grad student named Carl Willis had given him the necessary know-how. A buccaneering technician named Bill Brinsmead had pitched in the parts to put it together. Philanthropists Jan and Bob Davidson had noticed that the nation's brightest kids were falling through the cracks, and opened a school to cultivate their unique talents. And last but not least, Taylor's infinitely patient parents, Kenneth and Tiffany, hadn't stood in his way, even though every instinct they had as parents screamed they should slam on the brakes.

All of these people were willing to set their skepticism aside and believe that a kid could do amazing things if given a chance. And after hearing his story, I believed Taylor could, too. He had tons of people rooting for him. We had extended our trust and placed our bets.

Now all Taylor had to do to prove us right was win.

THE JUNKYARD GENIUS

You already possess everything necessary to become great.
—NATIVE AMERICAN PROVERB

In Arizona, deep within the heart of the Navajo Indian reservation, there was a legendary science fair project made out of a 1967 Pontiac radiator, a sheet of Plexiglas, and sixty-nine soda cans. A thirteen-year-old boy named Garrett Yazzie had pieced it together. In 2005, this puzzling collection of castoffs made its debut at a science fair in Phoenix. Judges stopped, stared, and asked the obvious question.

"What is it?"

In a shy, halting voice that suggested he wasn't used to being the center of attention, Garrett replied that it was a solar-powered room and water heater. He was dressed in a traditional Navajo ribbon shirt, a handwoven wool blanket draped over one shoulder, and a head scarf wrapped around his shortly cropped black hair. This was Garrett's first year competing in science fairs, and one of only a handful of trips he'd taken off the reservation. Garrett had hauled his project on a bus over three hundred miles to enter this fair.

Garrett was clearly nervous, but he could nonetheless articulate with precision how his solar-powered heater worked: how sunlight streamed through the Plexiglas pane to heat the radiator, which was lined with soda cans painted black, to maximize the absorption of light. From there, the warm air could be piped into a room and raise the temperature by forty-five degrees, or heat water to a near-boiling two hundred degrees. The judges were impressed. They had just one more question for Garrett.

"So what inspired you to create this device?"

Garrett's answer was so surprising, it would be told and retold until rumors of his exploits would reach me. As a seventh grader, Garrett was too young to qualify for the Intel International Science and Engineering Fair 2005, which was just open to high schoolers. Garrett wasn't like most science fair kids, who typically hailed from top schools and privileged backgrounds. Garrett was from the Rez. His project was a pile of junk. So how did he end up becoming one of the most famous science fair stars of all?

For Garrett, I'd learn, necessity was the mother of invention. He didn't wake up *wanting* to make a science fair project. He just had to. Only once I hopped in a car and drove to where Garrett lived did I understand why.

Piñon, Arizona, is surrounded by some of the most breathtaking landscapes on the planet. Up north, the wind-sculpted buttes of Monument Valley rise high in the sky, while down south, the Painted Desert ripples in rainbow colors. To the east and west, the Grand Canyon and the Canyon de Chelly cut deep into marbled layers of earth. I could see why Navajo tribes passing through six hundred years earlier had decided to settle here and make this their home.

With wonders like these next door, any small town is bound

to look insignificant. Piñon, though, does a pitifully good job at it. Lying on a hot, flat expanse of sand, it has one supermarket, one gas station, three stop signs, and that's about it. There's so little traffic, no one bothers building fences, so cows amble freely, grazing on yellow patches of grass. As I drove along the dirt roads lined with dilapidated houses, the statistics I'd seen about this town rang true: More than half of Piñon's 1,190 families lived below the poverty line, with the average income amounting to $6,045 per person per year. Back in 2005, Garrett's family was no exception.

Garrett lived in a white, singlewide trailer with his mother, Georgia; two sisters, Gwendolyn and Geralene; and eighteen-year-old Geralene's two young boys. For a family this size, these quarters were cramped, with bedrooms and even beds shared. But far more troubling was the state of the trailer itself. From end to end, holes riddled the roof, walls, and floor, allowing leaks and drafts to seep in. When gusts of wind threatened to tear the roof right off, the family piled tires on top and prayed it would stay put. For running water, a hose snaked through the window. For a bathroom, they used an outhouse.

During winter, when temperatures dipped below freezing, the Yazzie family faced even more serious problems. The trailer's sole source of heat, a coal-burning stove, was not only expensive to run, it triggered asthma attacks in Garrett's younger sister, Gwendolyn. The family tried to make do with blankets, but oftentimes their need for heat or food would win out. Garrett couldn't remember a time when he hadn't heard Gwendolyn cough and wheeze through the night. Every few weeks, her breathing became so impaired that they'd have to rush her to the hospital, a full hour's drive away. In the backseat, Garrett held his sister in his arms, watching her gasp for air like a fish on land. When

Gwendolyn was ten, she nearly died. It was the most frightening night of Garrett's life.

Georgia was born in Piñon, but had moved off the reservation to attend a few semesters in college and get a job in construction. The pay was great, but soon after Garrett was born, Georgia could hear her homeland call to her. Many of her friends said she was crazy for moving back. In Piñon, jobs were scarce, plus Gwendolyn's long hauls to the hospital could prove fatal. But Georgia felt that being back on the reservation would be good for her kids in ways that were hard to quantify. As a Navajo, she'd been raised to believe that their Creator had placed them between the Four Sacred Mountains that served roughly as the borders to their native land: Blanca Peak in Colorado, Mount Taylor in New Mexico, the San Francisco Peaks in Arizona, and Hesperus Peak in Colorado. Outside these four peaks, Navajos left the protection of their Creator, and risked losing their sense of self. Georgia didn't want to see that happen to her kids.

Garrett's father, whom Georgia had met while working construction, moved out soon after Garrett was born, handing Georgia a blanket and a package of diapers and wishing her luck. Garrett had met his dad only once, when he was twelve and over at a relative's house. A man he didn't recognize shook his hand, told him he was his dad, and left it at that. Garrett didn't know what to say. Occasionally people told him he looked like his dad, but he honestly didn't want anything to do with him. *He had his chance to raise me and he didn't,* Garrett thought. His mom was his mother *and* his father, and more than enough.

In an effort to remain close to her kids, Georgia worked as a teacher's aide at the local school. It didn't pay well, so she was constantly worried about making ends meet. She tried to hide these fears from her kids, but Garrett knew better. One time, sensing that she was at her wit's end, he followed her into her bedroom, sat down on her bed, and extended a pinky finger.

"Pinky promise," Garrett said. "Everything will be all right. One of these days, I'm going to take care of you and the whole family."

Pinky promises were one thing; putting those words into action was another. At thirteen, Garrett decided it was time to grow up and become man of the house.

Garrett was sitting in seventh-grade science class when his teacher announced that in March, the American Indian Science and Engineering Fair (AISEF) would be held in Chandler, Arizona. Would anyone like to participate?

At first, no one raised a hand. Not that this was all that surprising. Piñon's public schools, which were poorly funded and plagued by shortages in everything from textbooks to paper, didn't exactly inspire the town's students to excel. Many of its high schoolers never graduated. Far fewer pursued college degrees, opting instead to work in construction or the nearby coal mines. Science fairs, in this context, seemed like a waste of time.

But within the sea of blank, bored faces, a lightbulb had quietly gone off in Garrett, all thanks to one word he'd heard: Chandler. Chandler, a five-hour drive from Piñon, was off the reservation—a full-fledged city complete with skyscrapers, restaurants, golf courses, and even a theme park. Aside from the occasional two-hour trip to Wal-Mart, Garrett had never had a chance to go sightseeing on the other side. If creating a science fair project was his ticket, then so be it. He raised his hand.

Before this point, Garrett had expressed little interest in science, or school at all. As for grades, he slid by on C's. He frequently skipped class. Nearly losing Gwendolyn, though, had changed him. Now, every day after school, he clocked hours on the computer in his mother's classroom, searching for ways to heat his home without coal. The options in Piñon were limited, but one thing Garrett had noticed, looking up at the cloudless blue sky day after day, was that Piñon had plenty of sunlight. For centuries, the Navajo had danced in honor of the sun, *shani diin,* so that crops would be plentiful. Perhaps Garrett could harvest the gifts of the sun in his own way.

One day, while he was discussing these ideas with his mother in her classroom, their conversation caught the attention of the special ed teacher—a big, bearded man named Doug Davis. Doug, a former science teacher and wrestling coach, had once taught at a rough inner-city school in Little Rock, Arkansas, but was forced to leave after standing up to a hulking boy who'd started slapping around girls in class. In Piñon, Navajos called him a *belagana,* which translates as "little white apple," referring to white people's tendency to turn red in the desert heat. In spite of this lukewarm welcome, Doug fell in love with the Navajo and their customs, odd though they seemed. For one, the Navajo never pointed with their fingers. Instead, they pointed with their lips, puckering this way and that when giving directions. If a coyote crossed their path, they turned back, or else risked ending up in an accident. They never combed their hair at night or whistled in the dark, since this could attract skinwalkers—evil spirits, dressed in animal skins, that could steal your soul. *Hogwash,* Doug thought. Then one night, while driving down a lonely stretch of road, he saw a strange, headless animal float before his headlights. It made his hair stand on end. Since then, he'd stopped whistling.

Doug had grown close to Georgia, and often volunteered to help with her kids. He took Garrett fishing and gave him a man-to-man talking to when he started playing hooky from school. Pleasantly surprised to see Garrett finally applying himself, Doug said he'd be happy to help Garrett with his science fair project and agreed that a solar-powered device would be a good way to go, given that "green projects" were all the rage these days. From then on, he fielded Garrett's questions. Within weeks, they'd drawn up a blueprint of a device Garrett could build on his own. Now all he needed were the parts. Only where, in a desolate place like Piñon, could he get his hands on anything? Garrett, it turned out, had a good idea of just where to go.

In Piñon, when cars and trucks break down, they typically get pushed off the road, into a field, and left to rust and gather weeds. Since most of the town's residents can't afford to call a tow truck, dozens of vehicles in various states of disrepair lie scattered across the countryside like large gravestones. Garrett had grown up riding his BMX bike through this surreal landscape, picking through the remains along with other kids, building ramps and flying over the edge. Here, he thought, surveying the wreckage, was where he'd find what he needed to bring his science fair project to life.

Ever since Garrett was a young boy, he'd known that the warmest place during winter was not his home, but a running car. The reason, he'd learned while hanging out at his cousin's garage, was a car's radiator, which keeps the engine cool by circulating liquid through a network of pipes to absorb the excess heat. Driving down a highway, a car's radiator disperses enough heat to warm two average-sized houses. This got Garrett wondering whether a radiator could be rejiggered to harness heat from a different source. Like the sun.

Garrett visited seven trash heaps and opened countless car hoods before he found a radiator in decent condition. It was from a 1967 Pontiac so caked with rust, it was impossible to ascertain the car's model. The radiator was far from pristine, but as Garrett pried it loose and shook off the dirt, he saw it had promise. Dragging it home, along the way he picked up other things he knew he needed, including plywood and a plastic funnel. Unable to find a hose, he cut an inner tube off one of his BMX bike wheels and used that instead.

Once he'd collected all the necessary parts, Garrett moved on to the next stage: construction. He hammered together a shallow two-by-three-foot box, then placed the radiator inside. He spray-painted everything black, then topped it off with a sheet of Plexiglas bought at a hardware store in Gallup, a three-hour drive from his home. After a few more days of tinkering, Garrett's project was finished. All that remained was finding out whether it worked.

Garrett hooked up the inner tube to his home's water supply, sat back, and waited. Within five minutes, he placed his hand under the spigot, praying the water would be warm. It wasn't just warm, but hot. Piping hot. That night, Georgia washed the dishes after dinner in warm water. The Yazzies weren't the type to loudly celebrate Garrett's success, although Georgia did tell her son that she was proud of him. Garrett didn't really care if his solar-powered heater would win anything at the upcoming science fair in Chandler. All that mattered was that it worked.

Still, Garrett thought, *winning would be nice.*

Today, solar power charges everything from handheld calculators to satellites in space, and interest in this form of alternative energy is growing. According to research by Worldwatch, seven states in the U.S. Southwest could provide ten times the amount of electricity used by the country today. The reason this hasn't happened yet is that construction is expensive. Garrett, however, had changed the rules. He had proved that, in a pinch, it was also possible to harvest solar power in cheaper ways, even for free.

While such an invention was an extraordinary feat for any seventh grader, it was even more remarkable coming from a kid like Garrett, who belongs to what is arguably the poorest—and most poorly educated—minority in America today. The statistics paint a grim picture: On standardized tests, Native Americans score worse than 79 percent of students in the United States. Nearly half drop out of high school, the highest rate of any ethnic group. Of those who graduate, one in five go to college, and only one in four of them finish. Add all these numbers together, and the odds that Garrett would one day obtain a college degree were about one in fifty.

Part of the problem is that reservation schools, unlike most state-funded public schools, depend solely on the support of the Federal Bureau of Indian Affairs (BIA), a division of the Department of the Interior. Technically, reservation schools are supposed to be funded at the same per-student rate as non-reservation schools, but typically they receive only half that amount. As a result, schools cram students into trailers and converted storage buildings for classes. Oftentimes, facilities are so poorly maintained that they haven't met fire or safety codes for years. At Little Wound School on the Pine Ridge Indian Reservation in South Dakota, a girl reportedly lost several fingers after they were crushed by a metal door that the school did not have the money to replace. In recent years, many of the nation's 187 Native American schools have been condemned and closed for good.

Lack of funding isn't the only reason Native American educational standards are behind the curve; another reason is cultural. As little as a century ago, school administrators assumed that if Native Americans were to survive and thrive, they'd have to assimilate into mainstream American culture. This belief gave rise to Indian boarding schools where white teachers gave Native American students English names, cut boys' braids, burned their traditional clothing, and administered beatings to anyone caught speaking the native language. Richard Pratt, founder of the first Indian boarding school, in Carlisle, Pennsylvania, set the tone in 1892 when he lectured a congress of educators: "Kill the Indian and save the man." Since then, many of these same schools have adopted the opposite approach and tried to instill an appreciation for Native American traditions in their students. Still, once these kids leave their tight-knit, cloistered communities for a college campus bustling with strangers, culture shock can ensue. Combine this with the fact that many arrive with little better than an eighth-grade level of education, and these students can easily get discouraged and drop out.

Phillip Huebner, the director of Arizona State University's American Indian Program, was all too familiar with these problems. Since his department was devoted to keeping ASU's thousand-plus Native American students from falling through the cracks, every day his office filled with kids who were struggling, frustrated, homesick, and on the verge of giving up. Phil knew that preparing these students *before* they reached college— back in high school and middle school—was a crucial first step. As a former science teacher, he also knew how science fairs could motivate kids in ways that just sitting in class and learning rote facts couldn't accomplish. In 2002, he combined these two observations in his latest outreach effort, the Arizona American Indian Science and Engineering Fair (AISEF).

During the fair's first year, more than 250 projects poured in, but their sophistication levels left a lot to be desired—kids were doing experiments like which flavor of gum lasts longest. To remedy this problem, Phil began training teachers on reservation schools on the nuances of scientific research, revealing how even simple experiments were complex if done right. To illustrate his point, Phil would describe an experiment testing the stickiness of three types of tape—packing tape, painters' tape, and duct tape—which is done by affixing strips of tape to the underside of a ruler then dropping pennies, one by one, into attached bags, one at the bottom of each strip (the stickier the tape, the more pennies the bag can hold before the tape detaches). Phil explained how even the tiniest, most trivial things mattered: that you should take care not to touch the tape on its sticky side lest this affect its stickiness; that when dropping the pennies, you should drop them from the same height and at the same time intervals; that the experiment must be repeated, again and again, to ensure that random chance wasn't responsible for the results.

Every year, Phil passed on this wisdom to tribal teachers, hoping it would in turn be passed on to their students. By the time AISEF was in its third year of existence, his efforts were paying off. More than five hundred projects rolled in from four states, and the quality had shot up significantly. In 2005, however, one project in particular caught his eye. Moving closer, he saw that the contraption was made out of a radiator from a 1967 Pontiac. Standing next to it was a very quiet, very serious seventh grader named Garrett Yazzie.

Phil wasn't judging the fair, but he had little doubt that Garrett would win. The judges at AISEF agreed, showering Garrett with awards, including first place. Phil noticed that while Garrett had barely cracked a smile during the entire competition, walking up on stage to receive his awards, he was grinning from ear to ear. Even then, though, it seemed like Garrett was trying to hide it.

Which made sense. The one thing Phil had learned and admired about Native Americans was their modesty. Unlike American culture, which encouraged individuals to hog the spotlight, Navajo society paid homage to other people, places, and things that had helped them achieve their goals. Like the sun. And the earth. In Phil's mind, these were the types of teachings we'd all be better off adopting. There might be some things Native Americans could learn from our culture, but we had a lot to learn from them, too.

Chandler, Arizona, was everything Garrett had dreamed it would be. The day after his victory at AISEF, he took one last look at the city's bustling streets and brightly lit storefronts before boarding the bus back to Piñon. But that's not where Garrett's sojourn out into the world would end. For the first time in his life, he realized he was good at something. He'd also heard there was another science fair coming up in a few weeks, in Phoenix. So he decided to build a bigger project: a solar-powered device that could heat air as well as water.

To get started, he asked teachers to save their soda cans for him, and he soon had a trash bag containing sixty-nine of them. Then he nailed together a five-by-five-foot plywood box and placed his radiator in the center and the cans along the sides. After spray-painting the whole thing black and topping it off with a piece of Plexiglas, he hooked it up to his home. Within hours, the air temperature within the trailer had risen by forty-five degrees Fahrenheit. The Yazzies went to sleep in a warm, quiet house. Gwendolyn's wheezing and coughing subsided. On the day of the science fair in Phoenix, Garrett hauled his solar heater back on a bus and returned home victorious a second time.

Sifting through his awards one evening, Garrett spotted an

application to yet another science fair he had qualified to apply for, called the Discovery Channel Young Science Challenge (DCYSC). Every year, more than seventy-five thousand middle schoolers across the nation applied to DCYSC in the hopes that they'd be selected as one of forty finalists, who were then flown to Washington, D.C., to vie for more than $100,000 in prizes. Garrett had never been to Washington, D.C. He had never been on a plane, period. It was an enticing carrot, but he figured there was no way they'd pick him. What business did he have, a kid who'd barely been off the Rez, mingling with some of the brightest kids across the United States?

"Still," Georgia and Doug pointed out, "why not try?" So every day after school, Garrett worked on his application. After turning it in, he started receiving updates on whether he'd made the latest cut. And he had—first to four thousand kids, then two thousand, then four hundred. In two more weeks, he'd find out if he'd made it to the final round.

Garrett started counting down the days. Fourteen. Thirteen. Twelve. On the final day, Georgia had never seen her son so eager to go to school. Once they reached his mother's classroom, they sat down at her computer, logged on, and scanned the list of forty finalists. Near the bottom, there he was: *Garrett Yazzie.*

His mother screamed so loud she scared him. Teachers flooded the classroom, their arms outstretched for a hug. Garrett, however, was somber as usual. "What's wrong?" Georgia asked, pulling her son aside. "Aren't you happy?"

Of course Garrett was happy. But he was now swept up in an entirely different emotion: fear. Washington, D.C., wasn't just a city. It was *the* city, his nation's capital. There, he'd be representing his school, his state, and his tribe. Most of the other thirty-nine finalists would come from private schools and privileged backgrounds. To top it off, while Garrett had often looked up at the sky and wondered what it would look like if he were on a plane, he was also afraid of heights.

"Don't go as a science fair student," Doug said encouragingly. "Just go as a Navajo. You are an expert on being a Navajo. *You* can teach *them*."

Days later, Garrett boarded his first plane.

Garrett peered out his plane window and did his best not to dwell on the fact that he was five miles from the ground. Georgia, sitting next to him, spent the majority of their trip nervously fiddling with her seat belt or gripping the armrests. Even after landing safely in Washington, D.C., Garrett still felt as if he were floating, through an alien landscape populated by towering marble statues and immense pillared buildings. At the fair itself, held in the National Academy of Sciences, Garrett noticed that all the other kids were wearing suits and ties. In his traditional Navajo ribbon shirt, headdress, and handwoven wool blanket, he drew more than his share of curious glances and questions.

"Do you still live in teepees?" one child asked. Garrett had to smile at that. "Yeah, I do," he joked. "A two-story teepee. There's another teepee for our garage." When a woman asked why his moccasins lacked rubber treads, he had another answer ready. "It's because the soles wore off during our Long Walk," he explained, referring to the American government's efforts in the mid-1800s to force Native Americans off their land onto new settlements. As Garrett introduced onlookers and judges to Navajo life, he also explained how his solar-powered heater adhered to Native American ideals of accepting the gifts that Mother Earth and Father Sky had to offer. It was a compelling presentation, but was it good enough to win? Garrett noticed that even the names of other kids' projects—e.g., "Effect of Antibodies to SDF-1 and CD 144 on the Incorporation of Adult Hematropoietic Stem Cells in Choroidal Neovasculature"—seemed to be written in a foreign language. As he looked around, it became painfully obvious to him that this was where his winning streak would end.

Garrett was right. At the awards ceremony, his name wasn't called when they announced first place. His name wasn't called for second or third place, either, although he was surprised to find he'd placed seventh. It was a huge accomplishment, but Garrett couldn't help feeling sad that his journey would soon be over. In the past six months, he had gotten glimpses of worlds he was grateful to see. Still, these were places he could visit but not stay in. Within days, he'd be back in his trailer in Piñon.

Then he met a man from General Motors named Michael Pierz.

Michael and his wife, Kathleen, were from Clarkston, Michigan. Their daughter, who was in prep school, had also qualified as one of the forty DCYSC finalists. Early on during their stay in D.C., Kathleen and Georgia had warmed up to each other and begun talking about their kids, as moms do. Kathleen was captivated by Garrett's story and told her husband about it that night. Michael, an engineer at General Motors, was intrigued. He'd never heard of radiators being used for solar power and was curious to learn more about the boy behind the idea. The Pierzes invited Garrett and Georgia to Clarkston for a visit. There, after getting to know the Yazzies better, the Pierzes hatched a plan. Their son, they explained, attended St. Mary's Preparatory School for Boys in Orchard Lake, a half-hour drive from Clarkston. Tuition was expensive, but the Pierzes would be happy to cover a portion of Garrett's costs and board him at their home. That is, of course, if Garrett wanted to go.

At first, the Pierzes' generosity left Garrett and Georgia speechless. Garrett also knew that if he moved to Clarkston, he'd be far away from his friends and family, in a different world. Georgia reminded her son that she'd ventured off the reservation and lived in Phoenix, and had done fine.

"If I did it," Georgia told Garrett encouragingly, "you can do it."

Garrett knew his mother was right. A radiator from a derelict '67 Pontiac had gotten him this far. He had to find out where it would take him next.

Before Garrett left Piñon for Clarkston, Michigan, a medicine man came to the Yazzie home. He took ashes out of a fire, spread cedar on top, and fanned the smoke toward Garrett as he and the medicine man prayed. Soon, Garrett would be leaving his homeland and the safety of the Four Sacred Mountains. The Navajos' Creator could no longer look after him. He'd be on his own.

In addition to packing clothes, Garrett took a small bag of corn pollen, which would help protect him during his journey. It soon came in handy when he and his host mother, Kathleen Pierz, were driving along a highway one day and a coyote crossed their path. "Could you stop the car?" Garrett asked. Then he hopped out and sprinkled a handful of corn pollen in the coyote's tracks and said a quick prayer. A few miles in front of them, they encountered a car accident. Kathleen turned to Garrett, flustered, and said, "I think you may have just saved our lives."

That fall, Garrett, wearing a button-down shirt and tie, set foot on the carefully manicured lawns of St. Mary's School. He saw green wherever he looked—green grass, green trees, green ivy winding up redbrick walls. Nearby, the parking lot gleamed with electric blue Porsches, mustard yellow Mini Coopers, and sleek black BMWs, testament to the school's wealthy staff and clientele. It was a different world from the sun-bleached, junk-filled plains of Piñon. Even the food was different. Garrett tried sushi, but promptly spit it out in a napkin. He tried Indian food, but found the spices didn't sit well with him. He missed his mom's mutton and fry bread. He missed home, period.

The Pierzes did their best to warmly welcome Garrett into their house. For the first time in his life, he had his own bedroom. Still, he couldn't help treating his new home like a hotel. He kept his clothes in his suitcase, rather than in the dresser drawers provided. He tried not to touch things, lest they break. The Pierzes' older son, who was a senior at St. Mary's, showed Garrett around his new school and taught him a few rules that would keep him out of trouble. One: The school's dress code was not negotiable. The top buttons of dress shirts must remain buttoned. Students must never wear white socks. Hair must never grow below the eyelashes, or cover the ears or shirt collar. Flout any of these rules, and you'd have to deal with Mr. Rychcik, the dean of discipline, who'd force Garrett to stand and stare at the wall for an hour, or write the school's honor code over and over.

During his first year at St. Mary's, as a high school freshman, Garrett struggled to keep up with the other students. Even with Mrs. Pierz tutoring him, it was hard going at the beginning. Determined to fit in, he joined the crew team, then lacrosse. He'd never played lacrosse, nor did he realize it was a Native American sport. "You should be good at this," his teammates said to tease him. Garrett wasn't, at least at first. One teammate named Craig, who'd been playing lacrosse since childhood, offered to practice with Garrett outside of school. As the days and months passed, they became best friends. Garrett improved at lacrosse, and at school in general. BMX biking, once his passion, became a distant memory. On his sixteenth birthday, Garrett celebrated by getting his learner's permit so he could drive Michael Pierz's Fiat in car shows. Eventually, at a dance held in conjunction with a neighboring girls' school, he met a girl he'd end up dating and taking to his junior prom.

During summers and holidays, Garrett flew back to Piñon. He was always glad to see his family, but as time passed, Georgia noticed that Garrett seemed restless and eager to get back. "I can't miss this flight, Mom, we have a rowing tournament," he once said. Kathleen Pierz, in an effort to help Georgia feel involved in her son's life, sent her photo albums filled with pictures of Garrett in front of ice sculptures at fancy restaurants, or going tobogganing with all his new friends. Georgia was happy to see that her son was acclimating to the outside world. Still, sometimes she wondered: Did Garrett like his new home more than his old one?

Every time he boarded a plane back to Clarkston, Michigan, Garrett felt guilty about leaving his family behind in their trailer. The roof still leaked. Drafts crept through cracks in the walls. They still used an outhouse. Garrett's life had changed dramatically, but for the rest of his family, pretty much everything had remained the same.

Then they got a visitor.

"Good morning, Yazzie family!"

Ty Pennington, the happy-go-lucky host of *Extreme Makeover: Home Edition,* was shouting through a megaphone in front of the Yazzies' dilapidated white trailer. Weeks earlier, news of Garrett's story had reached the producers of this hit television show, which builds houses for needy families. Now, leading a crew of carpenters and eight hundred volunteers, Ty announced that within one week, the Yazzies would have a new home—and not just any old home. Using Garrett's solar-powered heater as their inspiration, they'd decided to build the Yazzies a green home that would run completely off solar and wind power.

As Georgia, Garrett, Gwendolyn, Geralene, and her two young kids emerged from their trailer with their few possessions, they were informed that during the construction process, they'd be whisked away for a vacation at Disney World in Orlando, Florida. It was the last time they'd see their trailer standing. After they left, a backhoe trundled in. Within minutes, the trailer was torn apart. Over the next seven days, in its place rose two traditional hexagonal Navajo dwellings called hogans—one for Georgia and one for Geralene and her two young kids, connected by a shared entry. On the roof, instead of tiles, indigenous plants blanketed the surface and served as insulation. A state-of-the-art air filtration system was installed to help alleviate the last traces of Gwendolyn's asthma. A ceremonial fireplace was erected in the center of one hogan so Georgia could perform traditional Navajo ceremonies. Outside, a windmill and a bank of solar panels guaranteed the Yazzies would never have to worry about their electric bills ever again.

Arriving back in Piñon a week later, the Yazzies took one look at their new home and unanimously started crying. Inside the entrance, Garrett's solar-powered heater hung like a piece of art. "Hey, buddy!" Ty Pennington said to Garrett. "So this is your invention, and we thought it would be fitting that it would be right here in the entranceway, right when you come in."

Garrett's response was classic Garrett: an understatement. "It's nice," he said, before Ty continued his tour.

Days later, Garrett returned to Clarkston to resume his studies. But he'd be back. He was sure of it.

"So this is it," I said as I stood inside the entrance to the Yazzies' home, staring at the 1967 Pontiac radiator and sixty-nine soda cans that had started it all. "It's beautiful." After allowing me to snap a few photos, Garrett and Georgia ushered me into the living room, where I sat down to chat about what they were up to now and how much their lives had changed.

"We still get a lot of spectators, especially when Garrett is home, who want to tour the house," said Georgia, who's been startled numerous times by strangers peering through her windows. While their hexagonal, hogan-style home boasted many amenities—a central fireplace for Navajo rituals, exposed pine beams, a flat-screen TV, and a dishwasher—Georgia said she preferred to wash her dishes by hand and admitted that she occasionally missed the derelict trailer where she raised her kids. She had a lot of memories there. Good and bad, but mostly good.

By the time of my visit, Garrett was a senior at St. Mary's Preparatory School for Boys. "It was a big transition," he said. "I woke up at seven, came home at three, and had five hours of homework. I was so far behind the other kids." Although he'd chosen to focus on his studies rather than continue entering science fairs, his accomplishments had inspired other Navajo kids to follow in his footsteps. In 2006, two students from the Gila River Indian Community qualified for the Intel International Science and Engineering Fair with projects that found optimal ways to farm-raise fish. To encourage more Native Americans to pursue the sciences, Arizona State University created a scholarship offering financial assistance to Native Americans, called the Garrett Yazzie Rising Star Scholarship Fund.

Once Garrett graduates in June 2010, he plans to attend college and study engineering, then start a company that will build windmills and solar panels throughout Navajo Nation. Did Georgia ever worry that Garrett would be lured away from the reservation by better opportunities elsewhere? "Yes and no," Georgia said. When I asked her what Garrett disliked about the outside world, her answer was instantaneous. "The cost of living," she said. "All that hard work you do is just going back to where you got it."

Garrett, in other words, hadn't forgotten the lessons his homeland had taught him, about appreciating the gifts that Mother Earth and Father Sky had to offer. In the future, he hopes to make a name for Native Americans in more dramatic ways. During a visit to the Henry Ford Museum in Dearborn, Michigan, Garrett and Kathleen Pierz toured an exhibit of famous inventors' homes, such as the cabin where George Washington Carver was born. "This museum will need to make room for a hogan," Garrett informed Kathleen. "Because I'm going to be the first Navajo on the moon."

Garrett knows that setting foot on the moon is a stretch. Still, he is unfazed by long journeys. After all, during the Long Walk, the Navajos traveled three hundred miles by foot, away from their homeland, then walked three hundred miles back once they were allowed to return. Hundreds died during that trek. But Garrett can always return to Piñon. And he wants to. The rest of the world is nice, but it's no place like home.

"ME? LEPROSY?"

Fear cannot be without hope.
—BENEDICT DE BARUCH SPINOZA

It started with a bug bite. Elizabeth Blanchard, known as BB to her friends, was at a picnic, sitting on the lip of a fountain under some oaks. It was late spring in Baton Rouge, Louisiana, balmy but bearable in the shade, although everyone down here knew that sitting in the shade came with risks. Oaks were home to buck moth caterpillars, one of which had fallen from the boughs above to land right where BB was sitting. Once the back of her calf brushed up against the black, spiny insect, it didn't take long for it to employ its stingers—not just once, but three times.

BB screamed. Friends and family came running. They were sympathetic, but hardly surprised that BB was the one who'd been hit. BB, for some mysterious reason, was notoriously accident-prone. Once, while walking down the driveway, she'd tripped and sprained her ankle. Another time, while roughhousing with her older brother Andrew, she broke her collarbone. Cars had run over her glasses twice, plus she'd crushed them underfoot once herself. BB seemed sometimes to be a walking disaster. It was as if she were cursed. The bright side, if there was one, was that this constant string of calamities rarely rattled her. Long before the caterpillar-induced rash on her calf faded, she had forgotten about it.

Then the rash returned. "What's that thing on the back of your leg?" her friend Leah asked one day as they were walking to class, much as she might have asked *What's wrong with your hair? It looks horrible.* BB looked down below her plaid navy skirt, the required uniform at the all-girls' school where she was a freshman, to see three red, raised marks—the same place and pattern left by the caterpillar four years earlier. It was strange that they had resurfaced, but it didn't worry BB at first. At her annual physical, her pediatrician was also unconcerned. "A minor skin irritation," he explained, handing BB some lotion. "It'll clear up eventually."

The rash stuck. Then it started to spread, creeping up to BB's knee and down toward her foot. BB's mother, Anne, still didn't think much of it, but since she was driving Andrew to the dermatologist anyway, it couldn't hurt if BB tagged along. At the dermatology office, Dr. Courtney Murphy obligingly inspected BB's leg, assuming she'd promptly pronounce a run-of-the-mill case of eczema or psoriasis. But this rash struck her as unusual. To be on the safe side, she took a biopsy and sent it to the lab for analysis. The results would be ready in a couple of weeks.

Anne, a librarian, was busy putting the final touches on her annual book fair when she got the call. "I don't want to scare you," said her husband, Steve, who had just spoken with the doctors. "But they think it might be leprosy."

Leprosy? At first, Anne wasn't even upset. She was incredulous, as if she'd just been told her daughter had been abducted by aliens. "People still get leprosy?" were the first words Anne could formulate in response. Once Anne broke the news to her daughter, BB's mind flashed back to those cartoons she watched as a kid, about Jesus and the leper, huddled in a dark street corner, his face deformed, his limbs falling off. Alone. An outcast. Confronted with the prospect of how much her life might change, BB did what any sixteen-year-old girl would do: She burst into tears.

At the Intel International Science and Engineering Fair 2009, I fully expected to see a lot of strange stuff. Still, meeting a girl with leprosy was a surprise. What floored me even more was that BB Blanchard was the very antithesis of how I thought someone with leprosy would look and act. With her wavy brown hair, big eyes, and bubbly personality, BB was pretty and popular, and a boy magnet to boot. As she sat down with me for an interview and told me her story, it soon became clear that everything I thought I knew about leprosy was wrong, and she was about to set me straight.

Back when BB first heard she might have leprosy, in October 2007, she was as clueless as I was about what that meant. Desperate for answers, she ran to the one place she hoped could help her make sense of her situation: St. Aloysius Catholic Church. BB had been baptized at St. Aloysius and had been sitting in those pews every Sunday since birth. She had grown up believing in the power of prayer. Ask God a question, and he will answer. So she closed her eyes, stilled her mind, and asked. *Dear God, do I really have leprosy?*

To her surprise, God answered immediately. That, at least, is how she interpreted what happened next, as the priest ascended to the pulpit to read a passage from Luke 17:11. The story of Jesus and the Ten Lepers. As the priest droned on, BB could feel her face flush, as if she were sitting in the church pew naked. What were the odds the priest would pick this passage right when BB's fate hung in the balance? The timing of it struck her as ironic, even ominous.

"It's a sign," she announced to her parents that night. "It's a sign I have it."

A few days later, doctors confirmed that BB had leprosy.

At this point, religion offered her little comfort. Everything she knew about leprosy had come from the Bible, and all of it was bad. Sufferers were considered "unclean" and were required to ring a bell so others could scatter as they approached. It was less a disease than a divine punishment, although BB couldn't imagine what she'd done to deserve this. Deserving or not, the cards were dealt. The die was cast.

Still, it was at this point that BB did something unusual. She marched into her room, where the fuchsia walls surrounding her were filled with photos of her and her friends at parties, picnics, swim practice meets, and jazz dance class. Since BB was a rabid fan of Louisiana State University's baseball team, her most thrilling piece of wall décor was a prominently displayed autograph from Ryan Schimpf, which had been given to her by a guy named Gunter. BB attended an all-girls school called St. Joseph's, but she had plenty of guy friends. She'd even had plenty of boyfriends. BB cringed to think how much that might change now.

Curling up on her zebra-striped bedspread, BB opened her laptop and typed it in: *leprosy*. Knowledge, after all, was power. She would fight fear with facts. And the facts she found were light-years from what she'd expected. Leprosy even had a history right in Baton Rouge, where she lived.

On the night of November 30, 1894, a tugboat chugged up the Mississippi toward Baton Rouge, towing a coal barge carrying five men and two women. Every inch of towline had been extended between boat and barge, and for good reason: All seven of the barge's passengers had leprosy. They were being transported to Carville, the nation's first leper colony. While they had volunteered to become its first residents, many later occupants were kidnapped from their homes by public health officials and brought there in handcuffs. An eight-foot-high fence surrounded Carville to keep them on the premises. It was probably for the best, since neighbors threatened to shoot patients if they escaped.

Carville, back in its prime, had been a magnificent antebellum sugar plantation surrounded by ancient oaks. But by the time these seven residents arrived, it had been abandoned so long that the mansion was crawling with rats, bats, snakes, and vermin. The walls oozed moisture. Torrents of rain poured through the roof. Having little choice other than to make the best of it, they settled in. Aside from the occasional doctor's visit, they were largely left to fend for themselves. No carpenter, plasterer, or painter would dare visit. No nurses could be convinced to care for them. Then one day Carville's residents were surprised to find four nuns at their doorstep. The Daughters of Charity of St. Vincent de Paul had heard about their predicament and volunteered to help.

Bit by bit, the sisters and the patients fixed leaks, painted walls, and installed electricity and indoor plumbing. They raised chickens, grew fruits and vegetables, and beautified the landscape with flowers. Croquet matches became a popular pastime, as did movie night once they obtained a projector. The sisters, who were also registered nurses, set up a laboratory as well and began searching for a cure. In the 1940s, Carville hit upon a promising possibility with sulfone drugs. Under such treatment, distorted faces returned to their original shape. Pain and numbness in patients' limbs disappeared.

As their spirits rose, some of Carville's residents started sneaking out through a hole in the fence surrounding the plantation, to head to football games or hang out at a joint called the Red Rooster a half mile down the road. After hearing about the hole in the fence, the director of Carville enlarged it so they could slip through more easily, since by now there was little reason why people with leprosy couldn't, or shouldn't, live it up. While Americans at large had scant idea what was happening inside their country's first leper colony, in 1950 that started to change thanks to the memoir *Miracle at Carville*. Its author, Betty Martin, had been sent there as a nineteen-year-old New Orleans debutante. Within its walls, she fell in love with a fellow patient, snuck out under the fence, and married. Her book, which proved just how rich the lives of patients living with leprosy could be, became a *New York Times* bestseller.

Today, there's little need for a place like Carville. Only a handful of elderly patients still live there, since they consider it home. Part of the mansion was turned into a museum, while its research and treatment facilities were moved to Baton Rouge. Thanks to better treatment and detection methods, leprosy is rare, with only a half a million new cases diagnosed worldwide and 150 in the United States per year. While a lot remains unknown about the disease—scientists still can't explain how it's spread, and why certain people get it but not others—what is known is that it's treatable and not highly contagious. Ninety-five percent of people are naturally immune. Once treatment begins, the bacteria are rendered incommunicable and the disease can't spread. As testament to this fact, none of the staffers at Carville ever contracted the disease from a patient.

In spite of this good news, people with leprosy still face a difficult battle today. That's because while their prognosis has improved, the public's perception of the disease has barely budged since the biblical era. In an effort to remove some of the stigma, patient advocates have pushed to rename leprosy Hansen's disease in honor of the scientist who first identified the bacteria behind it. But a new name isn't enough. Just a few years ago, a man working at a Fortune 500 company lost his job and was flown out of state by escort once the CEO learned he had leprosy. The employee could have sued, but that would have entailed even more people knowing what he had, so he kept quiet.

Even doctors mistreat leprosy patients, hovering near doorways to avoid getting too close, or accepting paperwork and pens delicately between thumb and forefinger to avoid touching places the patient has touched. Since patients never know how people will react, many keep their diagnosis a secret from friends, fiancés, spouses, and children. Some sufferers, if single, remain celibate, even though sexual contact is considered completely safe. Even Betty Martin, the famed author of *Miracle at Carville,* was writing and living under an alias. Being honest was too great a risk.

Science had given doctors the tools to make leprosy a problem of the past. But science alone wasn't enough. Leprosy needed a new image. A poster child, perhaps. Someone young and pretty and popular to step forward and say *I have leprosy, and it's no big deal. Really.* Someone like BB, in fact.

Am I going to end up looking like that?

The patient's face looked like melted wax. BB, while walking through the National Hansen's Disease Center in Baton Rouge, had gotten a glimpse of a leprosy patient who was living proof of just how grim it could get. In addition to causing facial deformities, the bacteria behind leprosy, *Mycobacterium leprae,* also attacks the nerves in the hands and feet, leading to numbness that leaves these areas open to infection and amputation. As BB and her mother passed by patient after patient, many of whom were in wheelchairs, panic set in. Would this be BB's fate? Pam Bartlett, the clinic's social worker, quickly set them straight.

"What you saw out there is not your future," Dr. Bartlett said. Many of the patients at their clinic had begun treatment much later than BB. Since they had caught BB's infection early, a cocktail of two drugs—dapsone and rifampicin—should be able to keep her symptoms in check. If all went well, she wouldn't suffer any pain or disfigurement at all. As Dr. Bartlett walked them through the basics of what having leprosy would be like, BB could feel her fears ebb. Having leprosy wasn't all that bad. But even if she knew this, could she convince others? Her older brother Andrew's reaction gave her hope.

"Leprosy? Cool," Andrew said, barely breaking his stride down the hallway toward his room. Next, BB tested the waters by laying the news on her friend Catherine. Nothing ruffled Catherine, so BB assumed she would keep her cool about leprosy, too. No such luck.

"That's not funny," Catherine said at first; then, realizing it wasn't a joke, she freaked out. "Is it contagious?" she asked first, wondering if she'd caught it already herself. "Are you going to lose any body parts?" she asked next, envisioning pushing her friend around in a wheelchair. Next, BB shared the news with her friend Sarah, whose reaction wasn't much better. At the time, BB had made Catherine and Sarah swear to keep her diagnosis a secret. But now that BB had all the facts, she decided to make a bold move: She'd tell everybody. Opening her laptop, BB typed up an email to her ten closest friends.

Hey guys,
The doctors confirmed that I've got leprosy. It's not
that bad. Don't worry, you can't get it. I'm taking
medication and I'll be fine, really.
BB

Her email inbox was flooded with responses. *I'm sorry! I
hope you're okay! Does it hurt? Are you sure it's not contagious?*
BB emailed them back one by one. *I'm fine. It doesn't hurt. It's not
contagious and no, I'm not going to die.* To her relief, no one
seemed repulsed. Still, these were BB's close friends. Of course
they'd rally around her, or at least put on a good show of doing
so. But what would other kids think? At an all-girls school, a student's status could change on a dime. Making matters worse,
St. Joseph's was a *Catholic* all-girls school. BB would be standing
up to the Bible itself.

BB donned her school uniform and took a deep breath. *Help
me out here, God.* Then she plunged into the locker-lined hallways of St. Joseph's to receive her verdict.

"So did you hear about BB?"

As BB made her way to class, it soon became clear that news
of her diagnosis was spreading quickly. Some students confronted her with questions, while others merely shot her curious
glances. For a while, BB couldn't shake the feeling that kids were
sneaking peeks at the rash on her leg—her own personal mark
of Cain. BB blushed in embarrassment. Uncomfortable with the
attention, she started wearing black tights. Anne, spotting the
tights, assumed the worst.

"Is someone picking on you?" Anne asked. BB insisted that

no one had said anything. Still, Anne couldn't help wondering what secret thoughts and fears were hiding behind people's polite veneer of concern. At swim practice, as the rest of BB's teammates jumped in the pool, BB was told she couldn't join them. Even though she had a note from her doctor saying she could resume all her usual school activities, her paperwork didn't specifically say she could swim, and her coach wasn't taking any chances. The swim coach said it was standard protocol, but Anne wasn't entirely convinced it was that simple.

News of BB's diagnosis had also reached the school principal, Linda Harvison. While she was comfortable navigating her school through emergencies ranging from flu outbreaks to hurricane warnings, leprosy was so far out of left field, she had no idea what to do next. Linda did know that the father of one of her students, a man named George Karam, was an infectious disease specialist at the Louisiana State University School of Medicine. Hoping he might be able to shed some light on the subject, Linda called him.

George, it turned out, had also heard about BB, since she was good friends with his daughter Claire and also took dance classes together with her. They'd even had a sleepover in his home just days before BB's official diagnosis. George understood Linda's desire to keep her students safe. Still, as an infectious disease specialist, he also knew that it was important to protect those who were ill as well. Fear could be as contagious as any disease. What BB desperately needed now was some damage control.

George explained to Linda how in spite of widespread beliefs to the contrary, leprosy wasn't highly contagious. Ninety-five percent of people, after all, were naturally immune to the disease, and under treatment the bacteria were rendered incommunicable. Relieved, Linda spread the word to her staff. Once they learned the facts, most teachers were fine with it, although occasionally some sensitivity training was in order. In religion class one day, when the teacher started talking about leprosy in a disparaging tone, Catherine's hand shot up.

"Hold up, my best friend's a leper," Catherine pointed out, as other kids in class rallied behind her.

Slowly but surely, leprosy was coming out of the closet. Everywhere BB looked, she saw references to the disease, from the medical drama *House* to the police show *Monk*. Once, while out at a movie theater with her friends watching *The Dark Knight,* BB heard the Joker say to Batman, "They need you right now, but when they don't, they'll cast you out like a leper." BB used to cringe at the sound of the L-word, but by this point, she saw it as an opportunity to crack a joke.

"That's me!" BB piped up in the theater. The word *leper* was out of the bag. Among her friends, it started catching on as a term of affection. "Hey, leper!" they'd shout down the hallway at school. One pal even programmed BB's name in her cell phone as *leper* rather than BB. Around Halloween, Catherine's boyfriend suggested that BB dress up as a leper, with a robe and a bell. After learning that one side effect of her medication was that it turned her pee orange, four girls crammed into a bathroom stall to watch BB pee.

As her notoriety grew, the rash on her leg began to retreat. Within six months, doctors said BB would be cured. Yet while she could have easily put leprosy behind her at this point, her curiosity about her condition remained. "If anyone were going to get leprosy, it would be you," Catherine said, referring to BB's endearingly accident-prone nature. After visiting the museum at Carville and seeing just how bad it could have been and still was for many leprosy patients, BB started thinking. *Maybe there's a*

reason I got this. Maybe I'm meant to do more. During her sophomore year, that opportunity would arrive as she sat down in science class next to Caroline Hebert.

Caroline, a slim girl with straight blonde hair and an intense personality, wanted to be a dermatologist when she grew up. She
relished picking scabs and popping zits—her own and others',
if they'd let her. "Caroline, you're disgusting," BB would say, but
Caroline couldn't care less. She didn't merely take the news of
BB's leprosy in stride. She found it downright fascinating. Every
day, she would beg BB for a glimpse of her rash. Could she see it?
Could she touch it? When their science teacher announced that
they should start thinking about what to do for a science fair
project, Caroline pounced. Could she and BB please be partners?
Could they please do their project on leprosy? BB agreed. She'd
been thinking of doing her project on leprosy anyway, and it
would be even more fun to do with a friend.

BB even had a mentor in mind: Dr. David Scollard, the researcher who, a year earlier, had scrutinized BB's biopsies under
a microscope and made her initial diagnosis. BB and Dr. Scollard
had chatted briefly at a medical panel where BB had volunteered
along with other patients to talk about her experiences with leprosy. Most of the other panelists, BB noticed, were immigrants,
and at least in their fifties. One was missing a few fingers. That's
when it occurred to BB that her situation was unique. She was
the best-case scenario, the face of what it's like to have leprosy
today.

After the panel, Dr. Scollard told BB that if she ever needed
anything, he'd be happy to help. So when he got a call asking if
she and Caroline could come into his lab to do a science fair proj-

ect, he welcomed them in. Under his guidance, BB learned about leprosy inside and out. She learned that the only animals other than humans known to carry the disease were armadillos, which could help explain why Louisiana and Texas, with their high armadillo populations, had more leprosy diagnoses than other states. While BB had always suspected that she'd caught the disease from the caterpillar sting she'd received years earlier, Dr. Scollard said that this was highly unlikely. If anything, the caterpillar sting had saved BB, prompting her to notice her rash earlier than she might have otherwise.

For their project, BB and Caroline scrutinized tissue samples containing the pill-shaped bacteria *Mycobacterium leprae* under a microscope, tracking how the bacilli count dropped during treatment. They observed that even though medication killed all bacteria within two years, it could take up to five years for the body to eliminate the evidence. This had prompted some doctors to mistakenly extend treatment for years longer than necessary, possibly leading to side effects, such as liver damage. BB had avoided this fate. But many leprosy patients, lacking experienced doctors, weren't so lucky.

BB also learned that even today, many doctors throughout the United States were so inexperienced with leprosy that they diagnosed patients too late, after medication could keep its worst symptoms at bay. More troubling still, many doctors still considered the disease highly contagious and treated patients like pariahs as a result. BB was horrified. These humiliations could have easily been her own. They were happening to leprosy patients across the country. While tackling the introduction to her project, BB thought long and hard about what to write. Finally, she settled on full honesty: *The inspiration for this project came from fear, and then the thirst for knowledge. The first reason is as follows: On*

November 5, 2007, Elizabeth Blanchard was diagnosed with Hansen's disease.

At their school science fair, BB and Caroline talked to countless judges, doctors, parents, and students. Many did double takes when they learned that BB had leprosy. Judges were also impressed with their research on how the bacilli remained in the body even after the patient was cured. BB and Caroline advanced from their school science fair to regionals, then from regionals to state. There, they qualified for an all-expenses-paid trip to Reno, Nevada, where they would compete in the Intel International Science and Engineering Fair 2009.

BB wasn't merely making waves in the science fair world. Her story was also spreading far in the media world, with features in everything from the *Washington Post* to *People* magazine. Thrilled by the positive publicity, the National Hansen's Disease Clinical Center made photocopies of the *People* feature and started handing it out to new patients to show that leprosy was nothing to be ashamed of. After that, everywhere BB went, from dance recitals to the grocery store, she could overhear people murmur, "That girl looks familiar," or else they came right up to her and offered their congratulations. BB was the first outpatient in the history of the National Hansen's Disease Center to tell her story using her real name. That took guts, and it was a gamble that appeared to be paying off.

"Oh my gosh, BB, you're famous!" said Catherine enthusiastically. Soon, the Blanchards' phone was ringing from journalists across the country. *The Doctors*, a spin-off of *Dr. Phil,* begged to fly BB to Los Angeles to appear on their talk show. *Inside Edition* also called, prompting Anne to wonder whether all publicity was good publicity. "Don't do *Inside Edition*," a journalist from a local paper warned Anne. "They might throw in clips from *Ben-Hur.*"

Anne, figuring it was better to be safe than sorry, didn't return *Inside Edition*'s calls.

The most exciting call Anne received came from London, where television producers were doing a documentary about Father Damien, a Catholic priest who'd died of leprosy after working at a leper colony in Hawaii. As a result of his work, Father Damien was being canonized by the pope. Would BB and her family like to come to Rome, be filmed for the documentary, and meet the pope? For the rest of the day, Anne was walking on air. She and BB had been invited to Rome! To meet the pope! Her family wasn't cursed, but blessed.

Through it all, the Blanchards did their best to maintain their sense of humor. When a photojournalist snapping photos of BB asked if she had a boyfriend, her dad joked, "We don't have to worry about that. That's because I'm going to tell the boys, 'If you kiss her, you'll get leprosy, so you'd better keep your hands off her.'"

Back in middle school BB had had a number of boyfriends, but none since her diagnosis. Was leprosy to blame? She couldn't say for sure. At one point, when she had a crush on a guy, she confided in Catherine, "What if he thinks I'm a freak?"

"He won't," Catherine said. BB wasn't so sure.

The only way to override the awkwardness was to barrel through it. So that's what BB did. One Friday night, when she was out, her friends introduced her to a boy and nudged her to tell her story.

"I had leprosy," BB said nonchalantly to the boy. "I'm cured. It's fine."

The boy blinked. "Oh. Okay," he said. Then they resumed talking about other stuff.

BB's story wasn't a sob story. On the contrary, she had managed to turn a horrifying diagnosis into something almost cool. In a way, she felt chosen, plucked from this vast population for a unique challenge that forced her to push her limits and find out what she was capable of. If it weren't for leprosy, she would have never known just how strong she was, or how much her friends and family were willing to rally behind her. While she can't believe she'd ever say this, BB is almost glad she had leprosy.

"I'm not, like, mad that I got it," BB told me in summary. "I'm not jumping for joy. But it's kind of fun to say I'm part of the five percent of the world that can get this disease. And now I can educate people about it. And I did something cool with it. Like the science fair. It all kind of worked out."

I tried to wrap my brain around the fact that less than two years earlier, she'd been sitting at mass, praying to God: *Why leprosy? Why me?* Now BB could see a glimmer of the answer. At Intel ISEF, if she won, she could reach millions with her message. She could eradicate fear and replace it with facts. Maybe this had been God's plan for BB all along.

4

STARS BEHIND BARS

When it is dark enough, you can see the stars.
—PERSIAN PROVERB

The next stop on my quest to find past science fair stars took me to Eagle Point School. At least, it's *called* a school. But as I stared up at the ten-foot-high chain-link fence topped with razor wire surrounding the facility, it was clear to me that Eagle Point was a juvenile correctional facility, plain and simple. I parked my car and stepped up to the entrance. A steel gate clanged, then creaked open. Waiting to greet me were a few representatives from Arizona's Department of Corrections, who gave me a tour. All around us, packs of boys in yellow shirts and olive slacks ambled about and eyed me curiously. It was swelteringly hot, but women within Eagle Point weren't allowed to wear sleeveless shirts. I felt it best not to ask why.

At first glance, Eagle Point didn't strike me as the kind of place where science fair winners would flourish. Yet starting in 2007, a handful of boys incarcerated there had won top honors at fairs in Arizona. One boy won a full-ride scholarship to Arizona State University. Another qualified as a finalist for the Intel International Science and Engineering Fair 2008 in Atlanta, Georgia.

It was all due to a science teacher named Ken Zeigler. By the time I arrived at Eagle Point in the summer of 2009, Ken had resigned from the facility, although he did live nearby. "I can't go back there," Ken had said apologetically to me over the phone. "Too many bad memories. But please come by my house if you like."

As soon as I arrived at his front door, it became clear to me that Ken, in spite of his stately, square-jawed appearance, had a soft spot for things that need rescuing. "I warn you, this house is full of animals," he said as I tripped over a black cat and a large white rabbit, which had hopped through the front door one day and stayed. The Zeigler household was a magnet for strays.

His wife, Mary, handed me a plate with which to partake of their barbecue and told me the story of how she and Ken met in a college course twenty years earlier. Mary had showed up late, frazzled from dealing with her two kids in the wake of a divorce, her bleach-blonde hair disheveled. Somehow, Ken saw through that. On their first date, they went stargazing with Mary's daughter, who told her mother, "He's going to be my next daddy." Ken proposed to Mary three months later.

Before accepting his post at Eagle Point in 2007, Ken taught science at a school in Texas. He and Mary, who was also a teacher, had lived so close to the ocean they could smell the surf from their front door. It was a good life, an easy life, one that Ken could see living for a very long time. Then one day while visiting family in Arizona, he bumped shopping carts with an old work colleague named Dan Lanphar.

"I'm principal of a juvenile detention facility called Eagle Point," Dan told Ken, handing him his card. "Come work for me, and I promise that whatever salary you're making, I can pay you more. A whole lot more."

Ken accepted Dan's card, but had little intention of accepting his offer. Yet a few nights later, he had a dream that seemed so uncanny, it convinced him to reconsider. In the dream, he was in the middle of a crowd of young men who seemed lost. Most of them ignored Ken, but a few of the young men approached. "We need you," one of the young men said. "Please come," said another. Ken tried to tell himself it was just a dream. But he couldn't shake the feeling that this dream was different, that it was trying to tell him something important. Maybe life had become a little too easy. At this point, he and Mary had raised two kids who were out of the house and on their own. Might Eagle Point be pulling at Ken because he was ready for a new challenge? On a whim, he picked up the phone and called Dan.

"You know I'm an oddball," Ken warned. "You know I'm not going to do things the way they're usually done."

"I'm counting on it," Dan said. That night, Ken had a conversation with his wife that sealed the deal.

"I'll miss the beach," Mary said and sighed. Still, in Arizona, the visibility would be great for stargazing. Ken could show those kids at Eagle Point a side of the universe they'd never seen.

On his first day at the school, Ken made his way past layers of locked doors, security guards, and surveillance equipment. He was surprised to find that in spite of its foreboding exterior, Eagle Point tried its best to look like summer camp. Within the inner courtyard, the central sidewalk was lined with trees, albeit scrawny ones. Beyond them sat six squat cinder-block buildings named after animals—Deer, Bear, Elk—which housed about a hundred inmates as they slept.

Every morning, kids emerged from their quarters and made their way to two larger buildings out back. There, they took their seats and studied English, math, science, and every other course you might find in a typical school. If, for a moment, Ken forgot that layers of razor wire surrounded them, he could almost feel like he was back in Texas, teaching regular kids. But the kids at Eagle Point were nothing like the kids from Texas, and it didn't take long for Ken's new students to set him straight. Soon after taking their seats, they responded to his friendly overtures with their standard greeting.

"Fuck you, Zeigler."

At first, Ken tried to remain encouraging. "Let's see how long we can all go without swearing, shall we?" he'd say, a suggestion that was inevitably met with more swearing. Profanity, however, was the least of Ken's problems, a fact that hit home when he was told he had to conduct a "sharp count" at the end of each class. At Eagle Point, pencils, or sharps, were popular weapons, and as such had to be kept under close watch. If Ken found that any were missing, he'd have to pat down each student until they were found. As an added safety measure, Eagle Point's kids weren't allowed to wear belts, pants with zippers, or shoes with shoelaces, since all of these things could easily be used to choke, stab, or kill other kids, or Ken himself.

Given that there were a variety of gangs within Eagle Point, classroom seating was another tricky issue. The best policy, teachers were told, was to keep the South Side kids away from the West Side kids, and a Latino gang called the MS-13s away from the Aryan Brotherhood. Chairs within the classroom were made of rubber in case they were picked up and thrown; desks and tables were heavily weighted so they couldn't be picked up at all. Teachers were even told to keep an eye on how much toilet paper boys used when they went to the bathroom, in case one of them stuffed the entire roll in the toilet and clogged up the plumbing in an effort to disrupt class. These kids were geniuses when it came to causing trouble.

If Ken encountered any serious problems, he was to report them via a radio attached at his side, which would send a battalion of youth correctional officers, or YCOs, running into his classroom. In case he ever found himself in a situation where he'd have to fend for himself, Ken was required to take six weeks of self-defense courses at the Arizona Department of Corrections Training Academy. The first step to subduing an Eagle Point kid, Ken learned, was to order him to "take a knee," or kneel on the floor. If a kid refused, Ken was told to resort to the primary restraint technique, a patented method of restraining inmates by crooking your arms around theirs and pressing your hands in the middle of their back. Ken hoped he would never have to resort to the PRT, but he would end up using it twice. He even heard rumors that one teacher had developed heart palpitations due to the stress of working there, while another teacher felt so terrorized, he quit on his very first day. For many of these kids, Eagle Point was just a pit stop on the way to adult prison, an example of which, called Lewis, happened to be conveniently located on the other side of the highway. The geographic proximity said it all.

Still, Ken wasn't ready to quit just yet. There was something he wanted to show these kids first.

For class one day, Ken brought in a basketball and a bunch of small flags. "Outside," he ordered his students. As Ken led them into the courtyard, a few YCOs hovered nervously nearby. Students weren't typically allowed outside during classroom hours, but Ken was tired of the classroom. These kids needed a hands-on demonstration of what the universe around them was really like.

Once they reached the northeast corner of the courtyard, Ken planted the basketball on the ground. "This is the sun," he said. "We're going to build a model of our solar system, to scale." Ken took out a rolling measuring tape, handed it to a student, and asked him to measure out thirty-three feet, then plant a flag. Assuming every foot in Eagle Point represented a million miles in space, that flag would represent Mercury. Head out another thirty-two feet, and that would be Venus. By the time they reached Saturn, they were nearing the chain-link fence at the opposite corner of the facility.

"What about Pluto?" one kid asked.

Pluto was impossible, Ken informed them, because if a flag were planted on the right spot, it would be located more than a mile away.

"No shit!" the student said. Ken smiled. His efforts to convey the vastness of the universe had finally hit home. Hoping to build off this success, Ken lobbied Dan Lanphar to let him put together a much more ambitious field trip. On the big night, a handful of kids from Ken's class set foot outside their unit. It was dark. Typically kids weren't allowed out of their bunks at night, since this was considered a security risk. But as Ken had suspected, none of the kids ran anywhere. Instead, they stared at a five-foot-high telescope standing before them, pointed up at the stars.

"Look here," Ken said, welcoming one of the kids to take a look. "Ever seen a crater on the moon up close?"

Ken showed them meteor showers, the rings on Saturn, and even entire galaxies outside the Milky Way, like Andromeda. He showed them just how vast the universe really was, and within seconds of their peering into the telescope, something clicked. Later, in class, kids started raising their hands, asking questions. Ken even noticed that kids were swearing less, then not at all, after he asked them again to see how long they could go without it. The next time Ken heard a curse word, it was in an entirely different context. He had brought in a plasma globe to show how electricity works. As the ball whirred, zapped, and extended its glowing tendrils into the air, Ken could hear one student in class utter in awe under his breath:

"That is so fucking cool . . ."

The kids started ribbing the student while he held up his hands in protest. "Swearing *that* way doesn't count!" he insisted.

These kids, Ken sensed, could go far if given the right encouragement. Over time, if they stayed out of trouble, maybe a few of them could go on to college, maybe even study science.

"There's a science fair coming up this March in Phoenix," Ken announced. Participants would be allowed out of Eagle Point for the day. It was a risky endeavor, but just the kind of carrot these kids needed to apply themselves. "A science fair project is a lot of work," Ken warned. "Still, if you do well, you may win an award, or even a scholarship to college. Who'd like to give it a try?"

Up stepped the first kid who would put Eagle Point on the map. His name was Lloyd Jones.*

Of all the kids Ken met at Eagle Point, no one seemed more out of place than Lloyd Jones. This pale, gangly seventeen-year-old

---

* The names of Lloyd Jones and Ollie Rodriguez have been changed, since they were minors and unable to be reached for comment.

looked like he belonged at baseball camp rather than juvee. While most of the boys tried their best to look tough in order to keep from getting their ass kicked, Lloyd loped into class every day and promptly started kicking himself.

"I'm so stupid," Lloyd would groan. "I guess that's why I ended up in here, huh?"

Lloyd's crime, Ken gathered, had something to do with hacking into computers. At one point when Ken was struggling to access some electronic files, Lloyd piped up with a suggestion that immediately solved Ken's problem. "You're good at this," Ken remarked, eyebrow raised in mock suspicion. Lloyd smiled. He was always smiling. Unfortunately, at Eagle Point, having kids like Lloyd wander around was akin to having a golden retriever gambol about in a ring full of pit bulls. Lloyd didn't belong to a gang, and therefore he had no one watching his back. So far, he had avoided getting hurt, but how long would that last?

Science fair season came along just in time. Lloyd started sticking around after school to work on his project for as long as Ken could remain on the premises. When Ken remarked that Lloyd was putting a lot of time into his research, Lloyd pointed out the obvious.

"In here, all I have is time," he said with a wry grin.

In addition to the usual challenges students face when putting together a science fair project, Lloyd grappled with a few added challenges. For one, kids at Eagle Point weren't allowed to use scissors, which were crucial for cutting paper to fit on the cardboard tri-panels where students presented their work at the fairs. Eagle Point kids weren't allowed to use rubber cement, since it was an inhalant, or glassware, since it could be broken and used as a weapon. In spite of these constraints, Ken made sure Lloyd made the most out of what he did have. For one, they had Ken's telescope. And thanks to Ken's knowledge of Arizona State University's online library, they had access to millions of images and data from space. If Lloyd needed pieces of paper cut, Ken brought them home and cut them himself, then brought them back in.

Day after day, Lloyd perused data about Mars taken by the 2001 Mars Odyssey orbiter. Lloyd had joined thousands of astronomers on the ultimate quest: to find water beneath the planet's surface. Where there was water, there was the possibility of life, and a place where humans could settle. Only where were the best places to start drilling for proof? Lloyd pieced together a theory about craters called lobate craters, which had a splash pattern. This pattern suggested that when meteors first formed these craters, they had penetrated the water table on impact. Large craters with a little splash, Lloyd reasoned, suggested the water table was deep. Tiny craters with a big splash suggested the water table was near the surface. Based on his calculations comparing the size of the splash pattern to the size of the crater, Lloyd pinpointed a place in the northern hemisphere where the water table might be easy to access. Several months later, when NASA's Phoenix Mars Lander started drilling into the planet's surface, it found water just inches beneath the surface there, proving that Lloyd's predictions were right on the money.

After a while, a couple other kids asked Ken if they could stay late after school to work on their projects. They, like Lloyd, felt like outcasts at Eagle Point and were grateful for the calm and quiet in Ken's class. Outside, trouble was brewing. They could sense it coming, like dogs do storms. Finally, one afternoon, it hit.

At first, Ken could hear yelling. Then he saw his YCO bolt from his post at the door. As Ken and his students peeked nervously out the window, they could see kids running, yelling, throwing rocks, breaking windows, and climbing on the roofs of buildings.

Eventually, it occurred to Ken that he was witnessing his first prison riot. Two of Eagle Point's inmates were being deported back to Mexico, and the Mexican gang was enraged. Ken tried his radio, but it was dead. He wondered aloud what they should do. Would it be safer to stay put, or try to make their way toward the administration building, since it was more securely guarded? Lloyd immediately chimed in with his two cents.

"I'm not going out into *that*," Lloyd said. "It's a jungle out there!"

The other kids nodded and agreed. "Can't we just stay here and work on science projects?" another one asked.

That's exactly what they did. For hours, the riot outside their door raged on, while Ken's kids quietly wrestled with statistics or designed the layout for their boards. Finally, a YCO ran into Ken's classroom and stared at Ken and his kids as if they were crazy.

"What are you guys still doing in here?" the guard asked, wild-eyed.

For lack of a better explanation, Ken held up his radio and explained that it was dead. "Fine," the guard said. "But we have to get these kids back to their units, immediately."

Outside, guards were trying to control crowds of kids and failing miserably. Eventually a plan was hatched where all the guards, staff, and teachers would lock arms, march in on the kids, back them into a corner, and cuff them. The plan wasn't without risks, but unless they did something, this riot might only get worse.

Ken looked frantically over at his wife, Mary, who'd recently accepted a teaching position at Eagle Point, in spite of Ken's reservations. Mary had asthma and neurodegenerative diabetes, both of which had worsened in Arizona. Ken had pointed out that the stress of teaching at Eagle Point wouldn't be good for her, but

Mary had stood firm that it was something she wanted to do. "If I can handle the pain I'm in already, everything else is water off my back," she'd said, so Ken had relented. But watching his wife march on a riot would be too much.

"*I'll* march," Ken said. "But my wife is not marching. This is not negotiable." As arguments broke out among the teachers and guards about what to do, the riot raged on until Eagle Point called in reinforcements from across the street at Lewis. Kids were corralled back into their bunks. Some were punished in solitary confinement.

Months later, when her health worsened, Mary resigned. Eagle Point was no place for Mary. It was no place for Ken, either. He had made progress with a few students, but his sense of hope was backsliding.

Science fair, though, was right around the corner. Maybe one of his kids would win a scholarship and leave Eagle Point better off than he'd been going in. Ken wasn't the only one who wanted the boys to taste success. After hearing that the students would have to wear their prison clothes to the fair for lack of alternatives, the staff pitched in and returned from the mall with a new set of clothes for each student. Everyone at Eagle Point was rooting for them. Ken didn't like to think what would happen if they came back from Phoenix empty-handed.

"Eagle Point? Must be some new charter school," Phillip Huebner mused when he heard that four of its students would be attending the regional science fair in Phoenix. As the fair's director, Phil welcomed newcomers. Then he learned that these four kids would be arriving in shackles. With guards. Eagle Point wasn't just some new charter school, but a school behind bars. What if these kids tried to escape or caused trouble? Still, Phillip reasoned, it wasn't right to ban these kids because of what they *might* do. And so, after hashing out some security concerns with his staff, Phil gave Eagle Point the green light.

On March 20, 2007, a van arrived at the convention center and pulled into the parking lot out back, since the front was mobbed with kids and impossible to guard effectively. Before entering the convention hall, the Eagle Point kids' handcuffs were removed so they wouldn't embarrass them, bias the judges, or draw stares from nervous parents. Beneath their pants, the students wore leg braces that would lock if they tried to run. But they had no interest in running anywhere. They were there to win, and Ken had prepared them well. In spite of their limited resources, their boards were crisp, their data flawless. During judging, their presentations were precise and polite. One could even argue that these kids were better behaved than the rest of the competitors, perhaps because they knew that for them this was it, their one chance to shine. Only was it enough? At the awards ceremony, Ken, four kids, and eight guards took their seats and waited for the answer.

"First place in the Earth and Planetary Sciences Division: Lloyd Jones, Eagle Point School."

Eight guards leapt to their feet and cheered like proud parents as Lloyd loped up on stage to accept his award. In addition to being the first juvenile detention detainee to win a science fair, Lloyd would also receive a full-ride scholarship to Arizona State University, which would be waiting for him upon his release. Lloyd returned to Eagle Point a legend, having blazed a path that other kids were eager to follow. Next year, once science fair season rolled around, Ken was deluged with kids clamoring to do projects. Among them was Ollie Rodriguez.

. . .

Ollie Rodriguez was one tough, troubled kid. Cursed with a hair-trigger temper, he was easy to rile up and lure into scraps with other students. Ollie's mom was Hispanic, and Ollie believed his dad was black, although he couldn't say for sure. His entire life was filled with uncertainties like that.

Ken wasn't sure exactly what Ollie was in for, and didn't ask. But one thing he did know was that Eagle Point could be good for him. Ollie needed structure and encouragement, two things he clearly wasn't getting at home. So when Ken was notified that Ollie was up for release, he was the only teacher to vote against it, quoting Shakespeare on Ollie's release form: *When he is best, he is little worse than a man; and when he is worst, he is little better than a beast.* If they released him, Ken warned, Ollie would be back.

Eagle Point let Ollie go. Within weeks, he was arrested again. Ken knew this because Ollie stormed into his classroom, having heard what Ken had said about his prospects outside Eagle Point.

"Did you really write that?" Ollie asked, enraged. "Do you really think that about me?"

Ken looked him in the eye and said yes. Ollie let loose with a string of expletives. Still, Ken wasn't bullshitting him. Ollie had to respect that. And besides, Ken was right. Ollie was right back where he'd started, stuck at Eagle Point.

"Do you want to do something about it?" Ken asked.

Ollie eyed him suspiciously. "You want to help me?"

Ken said yes, provided Ollie listened to what he had to say. So Ollie listened. Instead of getting into fights with students, he tackled his science fair project, staying late after class. Ollie set out to explore the possibility of extraterrestrial life. It was an ambitious venture, but he gnawed away at the data available on the universe's 270 known planets, taking into account planet size, what the planet was made of (an iron core projected a magnetic field that would protect the planet's surface from radiation), orbit (a circular orbit provided a more stable environment than an elliptical one), and other factors. While many astronomers focused on planets for signs of possible life, Ollie homed in on moons, which a growing number of scientists had identified as more fertile territory. Based on his calculations, Ollie whittled his 270 options down to a short list of eight planets whose moons could conceivably support life, including 47 Ursae Majoris b, HD 208487 b, and HD 213240 b. Reviewing his work, Ken could see that Ollie wasn't just bright, but brilliant. Ken nicknamed him the Evil Genius, an epithet Ollie didn't mind one bit.

In between discussing habitable zones and the effects of planet size on gravity, Ken heard bits and pieces of Ollie's story. Most of his family belonged to a gang called Glendale. The way you joined was you allowed gang members to beat you. If you wanted out, you were beaten again. Fewer survived getting out than getting in.

Ken had also mentioned to his students that he was in his own gang.

"Really?" The kids perked up. "What's it called?"

"The Lions," Ken said.

"Never heard of them," the kids said. "What do they do?"

"They troll around town looking for old pairs of glasses," Ken said.

When Ken's students still didn't get it, Ken chuckled, then explained that the Lions were a volunteer group that recycled prescription lenses for needy kids. Ken's world was as alien to the students at Eagle Point as their world was to him. Slowly but surely, though, they were learning. Over time, Ollie became so transformed that Ollie's mother made a point of finding Ken and telling him so during one of her visits.

"You must be Mr. Zeigler," she said, shaking Ken's hand. "I appreciate what you're doing for my son." Ken was surprised. Given what he'd heard about Ollie's family, he assumed she would look more hardened. But she looked like any other mother: worried, harried, and grateful for his help.

On Ollie's fifteenth birthday, after pulling some strings with school administrators, Ken was allowed to give him a soda, a rarity at Eagle Point. Ken also gave him something he wasn't allowed to: a one-gigabyte USB drive. Science fair was around the corner, and Ollie desperately needed a way to store his data and keep it safe.

"For me?" Ollie asked, staring at the USB drive in disbelief. "This is mine?"

"Technically I'm not allowed to give you that," Ken said. "But for now, you can use it. And when you get out of here, just take it. It's yours."

Ollie's face cracked into a huge grin. "Thank you, Mr. Zeigler." While most of Ken's kids called him *Ziegler*, Ollie called him *Mr. Zeigler*. This small effort to show respect was just one of many changes Ollie was undergoing. Even before being put behind bars at Eagle Point, Ollie had always felt trapped. But through science, he was beginning to see a bigger world, beyond where he'd grown up. Maybe that's why Ollie loved studying the possibility of life on other planets. It was a mental escape light-years from his own life. Ollie wanted out. And the upcoming science fair in Phoenix could be his exit.

On March 18, 2008, a van from Eagle Point pulled up in back of the convention center, just as it had a year earlier. Out hopped three kids wearing leg braces and carrying tri-panels, flanked by one guard. At the awards ceremony, Ollie's research on extraterrestrial life won Grand Prize, as well as additional awards from the United States Army, Air Force, and Navy. Ollie was also awarded the most prestigious honor of all: an all-expenses-paid trip to the Intel International Science and Engineering Fair 2008 in Atlanta, Georgia.

At Intel ISEF, Ken knew, Ollie would have the opportunity to present his work to Nobel laureates and other distinguished guests, competing against more than fifteen hundred of the world's brightest kids for $4 million in prizes and scholarships. It was an extraordinary opportunity. But it was also a risk. At Intel ISEF, the crowds would rival those of a Super Bowl stadium. It would be a cinch for Ollie to slip away in the crowd.

Given that no juvenile detention facility had ever faced such a dilemma, Eagle Point's administrators convened to discuss what to do. Ken spoke passionately on Ollie's behalf. In less than a year, Ollie had transformed before his eyes from *little better than a beast* into a model student. Perhaps Intel ISEF was just what Ollie needed to clean up his act for good. Administrators listened, nodded, and decided Ken was right. It was worth the gamble. They even allowed Ollie to attend without a leg brace or restraints of any kind, escorted by Ken and two guards who kept a polite distance. Ollie was free. Now all they had to do was wait and see if Ollie would rise to the challenge.

Once they landed in Atlanta, Ollie plunged into the crowds filling the massive convention hall. At one point, Ken heard one of the guards sound the alarm he dreaded to hear.

"We've lost him," one guard said.

Ken was nervous, but he tried to consider the best-case scenario. "He probably just lost us by accident," he said. As the guards circled the convention hall scanning for the boy, Ken made his way to where he had a feeling Ollie would be: at his booth, with his project. There, Ken found Ollie chatting with a college professor. The guards saw Ollie, relaxed, then laughed.

"Guess we were worried for nothing, huh?"

Ollie didn't win an award at Intel ISEF. But what he did win was a new perspective. Ollie had ventured into an alternate universe filled with bright-eyed kids bound for college and promising careers. More shocking still, he was one of them. Ollie had what it took to leave behind his life of crime and gang warfare. All he needed to do was keep doing what he was doing.

Which is why Ken was mystified, upon returning to Eagle Point, to see Ollie unravel.

After Intel ISEF, Ollie reverted to his old ways. He slacked off in school, mouthed off to teachers, and clashed constantly with the other kids. Ollie never explained why. Maybe his peers were picking on him for turning into a nerd. Or maybe he just couldn't handle the pressure of trying to meet his potential. For whatever reason, Ollie's newfound confidence had flared bright but was fizzling fast, and there was nothing anyone could do to stop it. Not even Ken. It broke his heart.

Meanwhile Ken was dealing with his own problems. Recently, Eagle Point had undergone some staff changes and budget cuts that made him suspect that the facility's once-lofty aspirations were shrinking. Now the main message he heard was that he should keep kids busy. Busy with tests. Busy with GEDs. Busy, but not inspired. More and more, Ken felt like his unorthodox teaching methods were being watched, a hunch that proved to be true one day when he was called down to the principal's office. At this point, Dan Lanphar, the principal who'd hired Ken, had retired. The new principal was a congenial fellow, but he didn't have much tolerance for rule bending. That day, his concern was a videotape taken in Ken's class.

Inside each classroom at Eagle Point, there was a video camera to track the students. But it also helped administrators keep tabs on teachers. "We see that you allowed a student behind your desk," the principal said, showing a videotape to Ken where a student had leaned behind Ken's desk to point out something on his computer. Only at Eagle Point, kids weren't allowed behind teachers' desks. This posed a security risk.

"But this is a kid I trust," Ken said.

"That doesn't matter. If the rules are for one student, they must be for all."

Ken was written up. He left the principal's office feeling more frustrated than ever. All this student had been trying to do was show him a photo of the craters on Venus. Had Ken ordered him to step back, the kid would have been crushed. Ken had fought hard to break through to these students and convince them they could do something right. But every day, Eagle Point's rules conveyed they couldn't be trusted to do much at all. Ken didn't want to give in. But as the admonitions from the administration mounted, something told him his days at Eagle Point were numbered. So when a teaching position opened in a school nearby, Ken took it. The next time he bumped into Ollie, Ken could tell he knew the score.

"I hate to leave when you've got issues here," Ken said.

Ollie shrugged in a way that conveyed that he understood. "You've gotta do what's right for you, too."

"You need to stay out of trouble," Ken said. "Think before you act."

"I'll try," Ollie said. "Good-bye, Mr. Zeigler."

They shook hands. That was the last time they'd see each other. Ken knew that soon Ollie would be set loose again and would most likely flounder. He had even considered the possibility of taking Ollie into his home as a foster parent until the boy was strong enough to fend for himself. But Ken knew that would be overstepping the line. At some point, he had to let go. And it was time.

In the fall of 2009, Ken started teaching at Tolleson Union High, a town over from Buckeye. At first, it took him a while to adjust to the fact that he wasn't cussed out in class—bitterly, affectionately, or otherwise. Life got easier. Ken found more time to devote to his wife, Mary, and to writing his sci-fi book series *Tears of Heaven*. Still, he continued to push the boundaries of what school administrators thought was realistic. "I'd like to take the first group of high schoolers up into space," he suggested one day. As I took a tour of Ken's new classroom, the walls were plastered in posters with motivational sayings, like: *Shoot for the moon— even if you fail, you'll land among the stars.*

Ken's Stars Behind Bars program, the first and only program within a juvenile correctional facility to produce science fair winners, earned him Arizona's Teacher of the Year Award in 2008. In spite of Ken's efforts to put Eagle Point on the map, in February 2010, the facility closed due to budgetary constraints.

Ken never heard what happened to Ollie Rodriguez, although he heard from Lloyd Jones. After leaving Eagle Point, Lloyd enrolled at Arizona State College with his scholarship. But soon, he was struggling. The classes were too overwhelming, the transition too large a leap. Eventually he dropped out, but that didn't mean he gave up. Instead, he enrolled in courses at a community college. One day, he hopes to resume his scholarship at ASU, once he's ready.

Ken often wonders how his old students at Eagle Point are

doing. Occasionally, he stumbles upon some answers. Once when he was at the grocery store, a young man from Eagle Point nearly bowled Ken over to greet him, saying he was back in high school, working hard, staying out of trouble, and looking to enter science fairs in his area. Ken wrote down some suggestions, as well as his contact info.

Ken knows, now more than ever, that there's only so much he can do. For many of the kids he met at Eagle Point, there is no such thing as a happy ending. "I'd love to tell you this story's like *Stand and Deliver*," Ken said, referring to the 1988 film about a math teacher who teaches calculus to a bunch of wayward kids in Los Angeles. "But I couldn't have continued doing it. The environment was so spiritually caustic. There are times I wake up from horrible dreams about the place." Still, Ken holds on to hope that at least a few of Eagle Point's kids learned something about the universe we live in, and how it's much bigger than they'd ever imagined. That, at least, is what the kids at Eagle Point taught Ken.

HORSE THERAPY

It is not enough for a man to know how to ride;
he must know how to fall.

—MEXICAN PROVERB

The day Doc arrived, Katlin Hornig knew this one was going to be tough. Doc was a dark brown, twenty-two-hundred-pound Belgian draft horse. His hooves were the size of dinner plates. His left eye was opaque, blinded by an accident that had flowered into an infection. Doc's owners, who ran a horse farm nearby, had tried and tried to train Doc, with no luck.

"We can't drive him," they explained, which meant Doc couldn't be hitched to a wagon. "He's too easily spooked." As if to underscore this point, Doc was acting spooked as they spoke, prancing nervously around the holding pen. But Katlin, eyeing the horse up and down, saw potential. If anyone could get through to Doc, she could.

Katlin hoisted herself over the fence. Curious what would happen, the neighbors stuck around to watch. A crowd often formed when word got out that Katlin was taking a crack at training a new horse. To her neighbors, it seemed painfully clear that Doc had the advantage. Katlin was just a teenager, with sky blue eyes and long blonde hair tucked back under a baseball cap. She looked far too sweet to be bossing around an animal twenty times her size. But beneath her serene exterior, Katlin had a stubborn streak. It was a blessing and a curse. Today, though, it would work in her favor.

As Katlin slowly approached Doc, she summoned up one nugget of wisdom she'd learned through the years: *To train a horse, it has to trust you first.* Katlin noticed that Doc got jumpy when approached from his blind side, so she switched sides so he could see her coming. Katlin also knew you couldn't rush a horse. So she took her time. On the first day, all she did was trim his hooves. On the next, she bridled him up and ran him around the edge of the pen. Within three days, Katlin had coaxed Doc to engage in the holy grail of horse training: a game called Simon Says.

Horses, due to their herding instincts, copy the actions of other horses. Katlin learned this the hard way one day when she and a couple 4-H kids were leading three horses through a shallow, knee-high lake to give them a workout. The minute one particularly mischievous horse named Bubba got the bright idea of rolling onto his back and taking a bath, the other two horses fell like dominos, drenching Katlin in the process. After that, she learned to use this copycat effect to her advantage. Rather than stand by helplessly as horses copied one another, she convinced horses to do something far more astonishing: They copied her. If Katlin walked, Doc followed in her footsteps. If she stopped, Doc stopped. If she backed up, Doc backed up. Katlin's connection with certain horses was so strong that if she pretended to gallop, the horse galloped alongside her, a trick that typically made crowds gasp.

By the time Doc left Katlin's care a few weeks later, the once-skittish, half-blind horse was calmly pulling wagons behind him.

Katlin's neighbors were amazed. Word of her abilities traveled far and wide. Training horses, however, wasn't her only talent. She was also good at science, and had been competing in science fairs since first grade. As a testament to her success, blue ribbons from regional and state competitions adorned the walls in her bedroom, mixed in with ribbons from 4-H events and plenty of horse posters. But Katlin wouldn't truly hit it big in the science fair world until her senior year, when the seventeen-year-old would qualify as a finalist for the Intel International Science and Engineering Fair 2009. Her story would teach me that some of the very best science fair ideas sprang from simply looking at your own life and realizing, *Hey, maybe we're on to something here.*

Katlin's hypothesis was simple: Horses made her happy. But might horses have the power to make everyone happy? After all, therapy dogs were used in hospitals to improve patients' moods. Might horse therapy have some of the same healing properties? Katlin knew firsthand that horses had helped her get through her roughest moments, and even at her young age, she'd had a lot to get through. Starting, first and foremost, with her dad, Bruce.

Bruce Hornig, a burly man with a bushy beard, had a plastic tube snaking into his nose attached to an oxygen tank at his waist. Bruce suffered from pulmonary fibrosis and progressive respiratory failure. Only 38 percent of his lungs worked, which meant that every breath he took was actually a half breath. By age two, as soon as Katlin could walk, she was traversing the hallways of her local hospital, having memorized its layout by heart. That's because every day, she and her mom, Janet, would navigate their way to her dad's room, where she'd climb onto his bed, get a big bear hug, then talk about her day. The fact that her family gatherings occurred in a hospital didn't faze her. Then one day, when she ran down the hall and opened the door to his room, she saw something that sent her running right back out.

"Mom," Katlin said with as much seriousness as a two-year-old can muster, "that's not Dad's room." A different man was in this room, lying in her dad's bed. Then she heard this strange man say her name—*Katlin!*—in a tone that sounded simultaneously stern but amused. Katlin recognized that voice. After peering back through the doorway for a second look, it dawned on her that the strange man in the bed *was* her dad. He'd just lost all his hair. Katlin climbed into his lap and placed her tiny hands on his bald head, fascinated. As Janet joined them on the bed, the parents exchanged a look and decided it was time to explain.

"Dad is sick," they said. Three months after Katlin was born, Bruce had found a lump on his neck. Biopsies and blood work revealed that Bruce had Hodgkin's lymphoma, and that the cancer had already spread into his chest and stomach. After a lumpectomy and one year of chemo, Bruce's cancer went into remission, only to return three months later. After another year of chemo and another remission, the cancer returned again. At this point, doctors recommended a bone marrow transplant, which Bruce's health insurance didn't cover. "So in other words, you're telling me that I should just go ahead and die?" Bruce demanded of his insurer's customer service rep, who didn't know what to say. Bruce had reached a dead end. That's when Janet—"the brains of the family," Bruce always called her affectionately—stepped in.

With some digging, Janet discovered that Bruce could volunteer for an experimental bone marrow transplant sponsored by the National Institutes of Health. After undergoing the procedure and taking a cocktail of new drugs, Bruce remained cancer free, but at a cost. Seven years later, he developed pulmonary fibrosis and progressive respiratory failure. None of the drugs Bruce was on are used today due to these side effects. Of the fifteen study subjects in this clinical trial, Bruce was the only survivor.

Doctors couldn't explain why Bruce was still alive, but his wife, Janet, knew exactly why: Bruce was stubborn, plain and simple. They'd met back in high school, bumping into each other at the county fair. Bruce was showing his pigs, but as soon as he saw Janet, he refused to let her out of his sight. That evening he followed her home, where they sat on the back of his truck and talked until well past midnight. Bruce's parents chewed him out the next day for getting home so late, but Bruce couldn't have cared less.

A year after they'd graduated high school, Bruce proposed, and Janet accepted. Janet went to college, while Bruce, to make ends meet, got a trucking license and started hauling loads across state lines. After Katlin was born, the Hornigs planned on having more kids, but once Bruce's cancer kicked in, that was no longer an option. Katlin was it, and as far as Bruce was concerned, she was more than enough. No matter how exhausted he was, he'd put on his best game face whenever she was around. Bruce and Katlin became so inseparable that by the time she was six months old, the infant was riding in Bruce's semi during short trips. At age four, Katlin took her first shower at a truck stop— using a ticket to shower amused her to no end—and when the waitress at the diner asked her if she'd like some coffee, she accepted, and her dad didn't object. On the drive home, Bruce lowered the windows so she could feel the wind whipping through her long blonde hair. Bruce did his best to hide the resulting tangle of knots under a baseball cap, but the instant Janet laid eyes on their daughter, she turned to Bruce and sighed.

"You rode with the windows down, didn't you?"

It took Janet two bottles of detangling spray to undo the damage. When Janet gave birth to a daughter, this was not what she had expected. Typically, Janet took an hour to get ready to go anywhere, and took special care curling her shoulder-length brown hair. Later on, she would relish the few times she'd get to help her daughter dress up for a homecoming dance or the prom. Still, no amount of satin dresses or hair detangler could change the fact that Katlin was her father's daughter, through and through. When Katlin turned eight, this particular fate was sealed after Bruce put her in charge of her very first horse.

His name was Boom because he was born on the Fourth of July. The day he arrived, all two thousand pounds of him, on their front doorstep, Janet sighed yet again, then let it go. Seeing the look in her daughter's eyes, she knew it was hopeless. Katlin was going to train Boom or die trying.

The very first lesson Bruce taught Katlin about horses was that people didn't train horses. Horses trained people. "Any weakness you have, the horse will pick up on it and amplify it," he warned. If, for example, Katlin was overly emotional one day because her dad was criticizing her training techniques, the horse would also throw a fit, causing Katlin to burst into tears a second time and her dad to gripe, "Goldangit, why are you crying?" Realizing such outbursts got her nowhere, Katlin learned to rein in her emotions and take setbacks in stride. By age thirteen, she'd stopped crying completely. If Katlin was calm, the horse was calm. Easy.

Gretchen was a whole different story. By this point, Katlin prided herself on her abilities to catch a wild horse. Typically it would take her under an hour. But Gretchen had other ideas. Every time Katlin approached, Gretchen would gallop just out of reach. This charade continued, over and over, for an hour. Then two. Then eight. By the time Gretchen allowed herself to be caught, Katlin was exhausted and exasperated. At least, she assumed, things would go easier tomorrow. The next day, Gretchen took Katlin on a ten-hour chase. "See?" her dad said. "Gretchen is teaching you a lesson." The instant you assume things will be easy, he said, was the moment it all went to hell.

Horses, Katlin came to realize, were far smarter than most people gave them credit for. One horse, named Tillie, learned how to drink out of plastic water bottles, which came in handy when Katlin would take her to competitions where troughs of water weren't readily available. Another horse, named Bubba, quickly acquired the nickname Houdini due to his uncanny ability to undo door latches and escape from stalls or fenced-in enclosures. If there was trouble to be found, they assumed Bubba was to blame. Once, when they arrived home to see that their horses had broken free and wandered to the other side of the road, then stopped, having realized that the grass was no greener on the other side, Katlin and Bruce didn't even have to ask which horse could have led such a fiasco. It was Bubba, all the way.

As Katlin's love of horses grew, so did the size of their herd. "We've got five horses, what's one more?" was Bruce's reasoning. Between the two of them, they would come to own eight horses, and these weren't your typical riding horses, but two-thousand-pound Belgian draft horses, similar in size to the Clydesdales in Budweiser commercials. Taking care of them was a lot of work, so Katlin typically rolled out of bed before dawn to feed the horses, clean stalls, mend fences, and do whatever else needed to be done before she rushed off to school. Occasionally Janet scolded Katlin for dragging saddles and harnesses caked with dirt into the house, but Janet knew she was fighting a losing battle. Katlin could just as happily hose herself down outside as shower in the bathroom. For haircuts, Katlin used the horse shears. On Bruce's hair, Katlin used the horse clippers. Out on the farm doing chores side by side, father and daughter had become so tuned in to each other's thoughts that they rarely talked much. By the time one of them might have needed to say, *Hey, could you grab an extra shovel?* the other would inevitably be standing there, shovel in hand.

Still, while Bruce tried his best to keep up with his daughter, his health was slowly failing him. Once, while he and Katlin were packing bales of hay in the barn, he stopped and clenched his chest. Katlin, who'd never seen her dad so much as wince, looked over at him, alarmed.

"Dad? You okay?"

"Sit down. Let's talk."

They talked for an hour. Katlin, who was twelve and had just begun taking science classes, asked questions about Bruce's condition. He answered each question honestly, without pulling punches. His daughter was old enough to know.

"They say I could be gone in three months easily," Bruce said. "But I've always proved them wrong. My goal is to see you graduate high school. From there, my goal is to see you graduate college. Then I'll try to hold on until you get married."

Deep down, Katlin had always known her dad could die any day. But it never truly sank in until then. For a brief moment, they allowed themselves to feel the weight of it. Then they both got up and resumed packing bales of hay, side by side. That was the Hornig way: Buck up and keep moving. It's a lesson that would soon come in handy.

As Katlin and Bruce trained horses and tried their best not to worry too much about what tomorrow might bring, inside the

Hornig household, Janet was worrying enough for both of them put together. Bruce's health wasn't the only problem plaguing her peace of mind. Given that his health insurance didn't cover many of the procedures for Hodgkin's lymphoma, Bruce's medical bills had ballooned to well over $250,000. Bruce's trucking had whittled this down, but by Katlin's freshman year in high school, his health had deteriorated to the point that he could no longer pass the physical needed to keep his commercial driver's license.

For Bruce, this was a humiliating blow. Being denied his livelihood was worse than the prospect of losing his life. He didn't care what happened to him, but he cared very much what happened to his family. To make ends meet, Bruce started building wagons and doing welding, while Janet took a job as an office manager. But it barely made a dent in the bills piling up. Now that Katlin was in high school, the question also loomed of how they would pay for college. Katlin wanted to be a veterinarian and had her heart set on going to Colorado State University in Fort Collins, which was ranked the second best veterinary school in the country behind Cornell. But how would they be able to afford it?

That's when the Hornigs hatched a plan: Why not just move there? While picking up and leaving their hometown of LaPorte, Indiana, for Colorado purely for in-state tuition rates might seem to be a drastic move for most families, for the Hornigs, drastic moves were par for the course. As an added benefit, the dry air would also help Bruce breathe more easily. And so, as Katlin finished up her freshman year in high school, Bruce made arrangements to sell their farm, while Janet flew to Colorado to pick a house. Once she found one—a modest one-story modular home perched on forty acres of farmland in Monte Vista—the Hornigs packed up their belongings and drove west, dragging six horses in trailers behind them.

Monte Vista, which is Spanish for mountain view, was surrounded by snowcapped peaks. Nestled within the San Luis Valley, the Hornigs' new home was even more rural and remote than their old one. Katlin started her sophomore year at Sargent High School, which was located in the middle of a potato field. There were twenty-seven kids in her class. Though small, Sargent High boasted a reputation for producing science fair winners, and Katlin was determined to be one of them. Now all she needed was a great project.

At the time, Katlin was having a tough time training a horse in her herd named Ariel, who was far more flighty and unpredictable than her other horses. Katlin also knew that Ariel was left leg–dominant, which meant she preferred leaning on her left hind leg over her right. Katlin also noticed that her right leg–dominant horses—Sadie, Jasper, and Tillie—were much more docile. Perhaps, Katlin mused, there was a connection between a horse's dominant leg and its personality. Perhaps, in the same way that handedness contained clues to a human's personality—left-handed people, for example, tend to be less linear in their thinking and more creative—hoofedness had an impact on a horse's personality, and how easy it was to train.

To find out if her hypothesis was correct, Katlin started making the rounds to local farms, asking residents if she could use their horses as test subjects. Most warmly welcomed her in. Katlin compiled her data by directly observing each horse's behavior and questioning its owners. Given that a large sample size produced more credible results, Katlin examined more than two hundred horses, squeezing in rounds of testing between school, homework, and her chores on the farm. For months, to keep up Katlin crawled out of bed at three A.M. and conked out at midnight. Every time she was tempted to throw in the towel, she thought about her dad. He never complained. So neither did she.

Once Katlin had collected her data, she tackled writing it up. That's where Katlin's mom, Janet, stepped in to help. While Janet couldn't make heads or tails of the scientific terms sprinkled throughout Katlin's text, she still knew a well-structured sentence from a bad one, and she proofread the paper draft after draft. "I don't know what this means, but it's grammatically incorrect," she'd say, pointing to a sentence. Janet wasn't just a wordsmith, but she had a good eye for aesthetics as well. So she helped Katlin put together her board, helping her decide which colors she should use in the background (dark blue and brown, she advised) and how to arrange the many pie charts and photographs so everything looked perfect.

By the time science fair rolled around her sophomore year, Katlin had amassed a wealth of research that found that right leg–dominant horses were indeed more rational and easier to train. Left leg–dominant horses were more skittish, unpredictable, and required a more patient approach, and therefore weren't ideal for new owners. Her findings won her first place at Colorado's state science fair and fourth place at the Intel International Science and Engineering Fair. Scholarship money started rolling in. Kids in Katlin's 4-H club nicknamed her Science Guru. One of the girls, inspired by Katlin's research, did her own science fair project on horses. The girl had heard that swirls—areas on a horse's body where the hair converges in a circular pattern—were correlated to certain personality traits. With Katlin's guidance, she started testing her hypothesis, hoping to one day follow in her friend's footsteps.

Another upshot of Katlin's research was that she and Bruce quickly gained a reputation around town as being horse people, so one evening when Bruce got a phone call from a neighbor saying a horse seemed to be suffering from abdominal pain, Bruce didn't hesitate. "We'll be right over," he said, as Katlin headed toward the truck carrying medical supplies. Based on the snippets she'd heard her dad discuss on the phone, she guessed that the horse probably had colic. Once they arrived at the farm to see the horse lying on its side, grunting in pain, Katlin knew her hunch was right. She and her dad both knew that the best way to rehabilitate a colicky horse was to keep it upright and walking, and administer a muscle relaxant. Only at this point, the seamless mental connection between father and daughter faltered.

"Did you grab the bandamine?" Bruce asked, turning to his daughter.

"I thought you grabbed it," Katlin replied, surprised, given how rarely these miscommunications between them happened. Bruce was already walking the horse, so Katlin turned back toward their truck. Since they were low on gas, the neighbors suggested that she take their truck instead. As Katlin hopped in and turned the ignition, she was so distracted that she didn't bother to fasten her seat belt. But it was unlikely that Katlin would pass many cars on these country roads. She'd be back in no time.

Katlin was right. She didn't pass many cars on her way home. But what she didn't see coming was the pothole, which would cause the truck's rear axle to snap in half. Which would send the car rolling off the road and Katlin flying through the window. Once she landed, the truck would land on top of her.

State patrolmen pronounced Katlin DOA: dead on arrival. Only one person begged to differ with that assessment, and that was

Bruce—who, after waiting a while, had decided to get in his own truck and find his daughter, suspecting that she must have gotten a flat tire on her way home. He arrived on the accident scene to see three-quarters of his daughter's body buried under three thousand pounds of twisted metal.

"She ain't dead," Bruce insisted, over and over, cradling his daughter's head, which was the only part of her body he could get his hands on. The cops quietly kept their distance, assuming Bruce had lost his mind. They gently tossed him a blanket to cover his daughter's body. From any objective standpoint, it was clear that the odds were heavily stacked against Katlin surviving the weight of a truck on top of her.

And yet what no one knew was that underneath Katlin, there was a ditch around eight inches deep and eighteen inches wide. It was exactly Katlin's size. Somehow Katlin's body had careened through the air so that she landed right in this ditch, like a puzzle piece in its place. No one would have believed Katlin's luck if she hadn't opened her eyes at that moment to see her dad hovering over her. He looked distraught, but she couldn't understand why.

"How's the horse?" she asked.

The cops, shocked, sheepishly called in an emergency helicopter. Bruce didn't even bother to tell the cops, *I told you she ain't dead, ya dimwits.* "The horse is fine," Bruce told his daughter. "Everything's going to be fine. Just hold on."

Katlin was flown to a hospital in Denver, where doctors confirmed that she had suffered a concussion and broken two vertebrae in her neck. Her face was a mess of lacerations, and her lips looked like raw hamburger. Thankfully, Katlin remembers very little, just brief flashes. Of doctors stitching her lip. Of her dad, sitting by her side the few times she regained consciousness.

Doctors warned Bruce that Katlin would be in the hospital

for at least two, maybe three months. But these doctors didn't know the Hornigs. By the third day, when transferring from the intensive care unit into another room, Katlin insisted on walking without a wheelchair—and carrying her own stuff. "The hell you are, Katlin" was her dad's response, but Katlin refused to budge until she had something to carry, so Bruce handed her a bouquet of flowers. After nine days, the doctor pulled Bruce aside. "I can't believe I'm saying this, but you and your daughter can go home," he said. Katlin's miraculous recovery baffled physicians, but Bruce wasn't all too surprised.

"She's got the Hornig stubbornness," he joked to Janet.

"I wonder where she gets it," Janet replied.

Katlin had survived the worst of it. Even then, it would take her another four months to fully recover. During the early weeks, Katlin couldn't lie down due to her neck brace, which made sleeping all but impossible. No matter what the hour or how tired he was, Bruce stayed up with Katlin to keep her company, falling asleep on the couch only after she did. Whenever Katlin had to remove her neck brace to change the dressings underneath, Bruce held her neck straight so it wouldn't wobble. Once Katlin was well enough to return to school, he carried her bag of books, like a high school sweetheart, and waited to take her home at the end of the day.

For years, Katlin had taken care of her dad. Now it was Bruce's time to return the favor. It was an odd role reversal, and at first it made Katlin uncomfortable. Since her neck was still healing, Katlin was ordered by doctors not to lift anything heavier than a gallon of milk. This caused a row every time she picked up a pitchfork to help her dad with the chores. Katlin and her dad also clashed over her desire to ride her horses again, which doctors warned would be suicidal if she did it too soon. If Katlin got bucked off a horse, she could end up paralyzed for life. So Bruce laid down the law.

"No, Katlin, you can't train your horses today," he said, no matter how much Katlin fumed. "You can go hang out with your horses, but that's just about it."

Katlin would storm off, march out to pasture, and call all eight of her horses at once.

"Tillie-Ariel-Bubba-Belle-Jasper-Sadie-Mocha-SheDaisy— HEY!"

Within seconds, eight horses came running. As they stood there, listening, Katlin poured her heart out.

For the first time in her life, Katlin's hands were tied. She couldn't ride. She couldn't work. All she could do was wait, and the waiting was driving her nuts. Due to her long recovery, she couldn't even do a science fair project during her junior year, and she had missed out on the scholarships that could have helped pay for college. Had she truly come all this way, and worked so hard, to see her dream of attending veterinary school go up in smoke?

Eventually, one of her horses would nudge Katlin with a velvety nose and remind her: *Hey, quit your fussing.* Even now, they were teaching Katlin a lesson, and it was one of the most basic tenets she'd learned through her years of horse training: patience. Not to mention perspective. Times weren't tough just for Katlin, but for people throughout the small town of Monte Vista. Even though it was still the kind of place where people kept their doors unlocked, it was proving increasingly unwise to do so. Burglaries were on the rise. Recently, there had been a rash of suicides. People were suffering, and unlike Katlin, they had no place to turn and talk about it.

That gave Katlin an idea. She headed down to the local police station in nearby Alamosa to talk to some cops.

. . .

Sergeant Ryan Black came from a long line of cops. His father, grandfather, and even great-grandfather were cops. All told, his family had seen a hundred years' worth of robberies, assaults, and murders in Colorado, and Ryan knew that seeing this side of humanity could take its toll. In his ten years on the force, many of his fellow officers had suffered from post-traumatic stress disorder. Two had committed suicide. And lately, things in their tiny town were getting worse. Crime was on the rise.

Then Ryan's boss gathered them together for a meeting. A high schooler named Katlin Hornig was looking for volunteers who'd be willing to try a little "horse therapy" for her science fair project. Most cops loathed all therapies—group therapy, psychotherapy, hypnotherapy—so Ryan wasn't surprised to hear his colleagues grumble that this new breed of therapy sounded way too warm and fuzzy. Ryan had another reason why he was reluctant to sign up. At age ten, he was thrown from a horse into a metal gate. Since then, he'd harbored a horse phobia and avoided them at all costs.

"But you ride a motorcycle, which is far more dangerous," Ryan's wife had pointed out countless times. Ryan had even crashed, twice, but kept riding. He didn't like being told what to do. Even though the captain of their police department preferred that the staff stay clean-shaven, Ryan wore a goatee expressly to piss off his boss.

Still, his wife continued chipping away at him. "A horse can't hurt you if you're not riding it," she pointed out. Meanwhile, Ryan's boss, a big horse man, gently strong-armed his staff until a few cops reluctantly stepped forward to help Katlin. For their first session, they drove down to a large open-air garage by the railroad tracks, where Katlin and her horses were waiting for them. The instant Ryan laid eyes on the two-thousand-pound draft horse, he stopped in his tracks.

"I thought you said you were bringing horses," Ryan said. "*That* is not a horse. That's a monster."

Katlin told Ryan not to worry. "Her name is Ariel," she said. "She's really nice. Let's just get you ready . . ."

Katlin took Ryan's blood pressure and had him fill out a form asking how he was feeling that day, which would help Katlin assess his mental state. Then Katlin led Ryan to Ariel and introduced them. Ryan stared at Ariel, unsure what he should do next. For lack of any other options, he began petting her, and talking.

"How are you?" Ryan ventured. "Do you like being petted?" Ariel's big brown eyes rolled toward him, which he took to mean that the answer was yes. As Ryan got more comfortable, he started cracking jokes. "At least you got me, instead of my buddy Red," he said. "Red would talk your ear off." That, at least, was what Ryan hoped. Recently, Ryan had been worried about his buddy Blain "Red" Shaffer. In the space of two months, Red had been the first on the scene of two suicides. Both of the victims had shot themselves in the head. Since then, Red, who typically talked so much that Ryan wanted to strangle him, had stopped talking. Red could act tough when duty called for it, but at heart, he was a softie. Red was also divorced, in his fifties, and didn't have a strong support network. Unless Red got some help, he was in trouble.

Out of the corner of his eye, Ryan was relieved to see that Red, who'd been paired with another horse, named Tillie, was indeed talking up a storm. Perhaps, Ryan mused, it was easier for a cop to pour his heart out to a horse rather than a human being.

"Would you like to give Ariel a massage?" Katlin asked Ryan, who happily obliged after she showed him how. As Ryan massaged Ariel's cheek, he was surprised to feel the weight of Ariel's chin on his shoulder. His massage had put his horse to sleep. During Ryan's next two sessions, Ariel conked out like clockwork every time he started stroking her cheek. *Man, I must give a pretty good massage,* Ryan thought.

After a half hour passed, Ryan had his blood pressure taken a second time and filled out more paperwork. Even without seeing the results, he had a hunch that his blood pressure had dropped, and his mood had improved immensely. Ryan felt fantastic. So did Red and the other cops who'd decided to give horse therapy a try. After that, cops were chomping at the bit for their next appointment. Katlin's therapy horses had won their first loyal fans. The sessions even improved Ryan's marriage, at least in a roundabout way. When he bragged about his talents at giving a massage to a horse, Ryan's wife had a ready retort.

"What, you're willing to give a horse a massage, but not me?"

Ryan now gives his wife massages regularly. Whenever he and his buddies in the police department hit a rough patch, they can call the Hornigs, who'll drive over with a trailer load of Belgian draft horses so the cops can pour their hearts out and return to work refreshed. Katlin also tested her horse therapy on individuals at a nearby correctional facility, one of whom got teary-eyed and insisted on having Katlin snap a photo of him with the horse to send to his family back home. After that, Katlin put her horses on call for anyone who needed them, from cops to correctional facilities, nursing homes to special needs kids. Everyone, Katlin found, could benefit from a little horse therapy.

. . .

Days, then weeks passed as the vertebrae in Katlin's neck continued to heal. After four months, doctors pronounced that she
had fully recovered and could hop back in the saddle, which Katlin did as soon as her father gave the green light. By science fair
season her senior year, she had also amassed a mountain of data
that proved beyond a doubt that her hypothesis was correct.
Horse therapy could significantly lower people's blood pressure,
reduce their stress levels, and improve their moods. In this light,
the Hornigs' resilience to all that life had thrown their way made
perfect sense.

Katlin entered her regional science fair, and won. She went
on to the state science fair and won there, too. That meant that
Katlin had qualified for the Intel International Science and Engineering Fair 2009, which would be held that year in Reno, Nevada, a fifteen-hour drive away.

Science fair kids rarely travel light. In the back of the Hornigs'
gray van, Bruce packed Katlin's science fair board, three suitcases
full of records and research, a duffel bag of clothes, and another
duffel full of schoolwork. Bruce would have killed to be on the
open road with his daughter again, just like back during his
trucking days. But a lot had changed since then. For one, they
had a farm and eight horses who needed looking after. So Bruce
volunteered to stay home and let Janet accompany their daughter. He hugged them both, told them to drive safe, then turned to
Katlin and uttered his signature send-off.

"Good luck, kick ass, and take names later."

Bruce watched as the gray van grew smaller in the distance.
He tried not to think of how much was riding on this trip. Between his medical bills and Katlin's own from the car accident, the

Hornigs didn't have the money to send their daughter to college. Winning a scholarship wouldn't just be nice. Without it, Katlin might not be able to go to college at all.

At Intel ISEF, Katlin would have a chance to win $4 million in prizes and scholarships. Only to do that, she'd have to beat more than fifteen hundred of the top science fair competitors from more than fifty countries. Many had access to laboratories that were far more sophisticated than what Katlin had on a farm. Bruce wasn't a scientist, or a doctor, or even a college graduate.

Nonetheless, he'd taught his daughter plenty.

"All my life my hero has been my dad," Katlin told me. "I've seen him go through tough times and excruciating pain and pull himself out of everything. So whenever I see myself hurting, I know I just have to keep going and everything will be fine." The fact that her dad could die any day doesn't faze her, at least not anymore. "When it's an everyday thing, you just deal with it," she said.

Later on, during a phone conversation with Bruce, I asked where he got his strength. The horses, he said, help immensely. "No matter how bad I feel, I have to get up and go take care of them," Bruce explained. "Even on my worst days, where my body says, *We ain't doin' nothin',* I've gotta go out and feed the horses." Bruce, like Katlin, had also noticed that talking to horses could have a therapeutic effect. "That's the amazing thing a horse can do for you," he said. "You can have the shittiest day, and you can just spill your guts. And when you're done spillin' your guts, they're still nudgin' you for an extra pet, saying, *Hey, I understand, I'm here, it's cool. Pet me behind my ears again.* And you feel so much better."

There was only one thing that haunted Bruce: He would

never rest easy until he knew his daughter would be okay without him. Maybe that's why he'd been holding on, breathing in half breaths, for all these years. Now all he could do was wait, and busy himself with work, and hope for the best. And Bruce had high hopes. After all, Katlin had eight huge horses pulling for her. That sure couldn't hurt.

6

THE KID WHO TOOK ON DUPONT

Size matters not.

—YODA, *STAR WARS*

Most people wouldn't consider Parkersburg, West Virginia, a hotbed of terrorist activity. Typically the biggest crisis in town might be a large snapping turtle stuck on the highway, backing up traffic. If there's one thing Parkersburg has plenty of, it's turtles, and water, since the town straddles the junction of the Ohio and Little Kanawha rivers. Turtle trouble, yes. Terrorists, no. Which is why Pete and Karen Welcker were surprised, on a sunny spring day in 2007, to find an FBI agent at their door.

"I have some questions," the FBI agent explained after the Welckers let him in. The agent's questions were about a blue minivan, and some photographs taken down at the DuPont chemical company. The Welckers answered these questions as best they could. But the individual who truly knew what was going on wasn't home. She was at school, working on her science fair project—the very thing that had brought the FBI to Parkersburg in the first place. The Welckers' daughter, who was seventeen and a senior in high school, hadn't talked much about what she was doing at all hours in her laboratory. But now that her name was etched in an FBI database as a terrorist suspect, it was official: Kelydra Welcker was in over her head.

After hearing rumors of this young Erin Brockovich, I set off in search of her. By now, I assumed, she was in college; with a name like Kelydra she shouldn't be hard to find on Facebook. I was right. In fact, she was the *only* Kelydra on Facebook. Her name, in a way, was what started her on this David and Goliath–type venture to begin with.

"I'm named after the snapping turtle, *Chelydra serpentina*," Kelydra chirped on the phone. Why, were her parents really into turtles? This question cracked Kelydra up. At home, in tanks stacked high in the kitchen, the Welckers had fifteen pet turtles. As avid members of Parkersburg's Reptile Rescue and Aid Society, they kept their eyes peeled for turtles in trouble wherever they went. Once, while driving past a forty-pound snapper waylaid in road construction, they screeched to a halt, hoisted the ornery animal into the trunk of their car, and ushered it into a nearby creek. At Chinese restaurants that served turtle soup, Kelydra's dad often bargained to buy every serving, alive and uncooked.

On weekends, the Welckers headed to one of Parkersburg's many waterside parks, like sandy Ravenswood or rocky River Road Run. Pulling her long brown hair back into a ponytail, Kelydra waded into the water alongside her older sister, Bonnie, and younger brother, Peter, to catch crayfish, dig for clams, and pluck snails off rocks to take home. Their mom, an insect enthusiast, classified dragonflies. Meanwhile their dad sat on the banks and waited for turtles. Kelydra was always impressed with how he could sit still for hours, patiently watching the water, the sun glinting off his white beard as if he were a statue of Santa Claus. Finally, Kelydra would see him nod, which meant a tiny turtle head had surfaced for a breath of air before dropping back under with a *bloop!* It might seem dull to most folks, but nothing thrilled Pete more.

Growing up in a family like this, with a turtle as her namesake, it was only a matter of time before Kelydra started saving turtles herself. At age six, she scared her parents by staging her first rescue, darting into the road where a semi was bearing down on a baby box turtle smaller than her hand. Rather than scolding her, the Welckers gave her a lesson in physics: "That truck was coming at you at seventy miles an hour. You, on the other hand, can only run three miles an hour. Get your space/time parameters straight before you get yourself killed." Kelydra, though, wouldn't have done things any differently. That tiny turtle needed her help. It was her duty, her calling, to fend for creatures unable to fend for themselves. Come science fair season, this belief would compel her to protect the place where turtles lived as well. The water.

Kelydra had competed in science fairs since kindergarten and was a science nerd through and through. She and her boyfriend, who was also a science nerd, swapped atrociously corny pickup lines over the phone just for fun. *I want to be your derivative so I can lie tangent to your curves. I want to be your helicase so I can unzip your genes.* Kelydra didn't care that science wasn't a particularly cool pursuit at her school, Parkersburg South. She had long ago given up trying to win any popularity contests. Science fair was the only contest she cared about, and she let nothing stand in her way, no matter how unappetizing the prospect. Once, for an experiment that involved raising mosquitoes, Kelydra shaved one of her gerbils with a razor, nicknamed him Shavy, and set him loose in a tankful of mosquitoes to see if the insects would bite. Her verdict: Mosquitoes didn't like gerbil blood, so Shavy escaped unscathed, grew his fur back, and lived to a ripe old age.

During her junior year, keen to come up with a winning science fair project, Kelydra spotted an article in the local paper that got her gears turning. PFOA (perfluorooctanoic acid), a chemical used by DuPont in the production of Teflon, had seeped into the groundwater surrounding the plant and shown up in local residents' blood tests. According to one study, the average American had five parts per billion of PFOA in their blood. People living near DuPont, like the Welckers, had on average six times that amount. In further studies, PFOA had been found to cause cancer in laboratory animals. While there was insufficient evidence to conclude that PFOA exposure caused adverse health effects in humans, tests were continuing, leading Kelydra to conclude that any living creature who drank, bathed, or swam in Parkersburg's water was doing so at its own risk.

After discovering a delay in DuPont reporting this issue, the Environmental Protection Agency slapped the company with a $10.25 million fine. DuPont volunteered to cut their PFOA emissions by 95 percent and installed carbon filters in nearby water treatment facilities, which would remove PFOA from the water. Curious whether those filters were living up to their claims, Kelydra decided to test local water supplies and present her findings at the school science fair.

During her first few months of testing, Kelydra found that the carbon filters were indeed doing the job. But four months in, she was surprised to see that her water samples' PFOA levels seemed to be rising. After alerting the water treatment facility to this fact, Kelydra learned that the carbon filters were most likely saturated and in need of replacement. By her estimates, given that carbon cost $1.25 per pound and the facilities' filters used thousands of pounds of carbon, replacing those filters every four months could cost millions of dollars over the years to keep the system working. And that merely took care of the water coming from the treatment facility. Families who got their drinking water from wells and cisterns, like the Welckers, weren't protected from PFOA at all. Kelydra was appalled. This just didn't seem right. But for the first time in her life, she was about to learn that right and wrong weren't the only factors she'd have to consider.

In Parkersburg, just about everyone worked at DuPont, and Kelydra's family was no exception. Her dad, Pete, had been a chemist there all his life, and, recently retired and in his early sixties, he depended on the company for his pension. Kelydra's older sister, Bonnie, was a chemical engineer there and was planning to marry soon and start a family. Kelydra was what's called a DuPont brat, born and raised on an unwritten set of rules that defined the company credo: *Do the best you can with what you've got. Do the assignment you're given. Don't ask too many questions. Don't discuss your work at home.* For one, your wife doesn't want to hear it, plus loose lips sink ships. Kelydra knew this was true because her dad said he'd once conducted chemical industry espionage, where he struck up a conversation with an employee at a rival company and innocently asked a few questions. Out came a piece of paper, on which a diagram was drawn, which Pete took back to DuPont and put to good use. Kelydra dreamed of one day following in her family's footsteps and becoming a chemical engineer at DuPont. DuPont, in a way, *was* her family.

All the more reason, as her dad saw it, not to bite the hand that feeds.

"I don't want you working on this project," Pete said, after hearing what his daughter had planned for the upcoming science fair. "Do you have any idea what you're up against? Kid, you're stepping into a league that plays hardball."

Kelydra knew that her dad had a point. He depended on Du-Pont for his pension, and the company had been good to him over the years. Even now, on any given weekend, Pete would head into town to catch up with his old coworkers, a habit Kelydra and her mom jokingly referred to as dad running his trap. By criticizing DuPont, Kelydra could be accused of trying to put the company out of business. Given that this was their home, their community, that wouldn't be right.

All her life, Kelydra had never heard her dad say no. Up until this point, she'd never given him a reason to. She was Daddy's little girl, who lived to make her father proud and had never gone through a rebellious stage. But now, at seventeen, she was about to join the ranks. Kelydra believed in what she was doing, and no one was going to stop her. Crossing her arms, she looked him in the eye and stood her ground.

"Tough luck," she said. "I'm not doing anything wrong. And besides, I'm just a kid. Why would DuPont worry about me?"

Pete, who knew his daughter well, could list a million reasons, starting with the laboratory Kelydra had built in their own backyard.

In Parkersburg, laboratories where science fair kids could conduct their research were nonexistent. The closest, at Ohio State, was a two-hour drive away, and space there was limited. Some students drove as far as Pittsburgh, which was three hours away. This posed a problem for Kelydra. The Welckers were a frugal family who ate casseroles rather than dining out. At Wal-Mart, they were horrified to find jeans *with rips in them* on sale for fifty bucks. Kelydra wouldn't dream of asking her parents to funnel so much cash toward gas traveling back and forth to Pittsburgh. So the solution to her problem was obvious. She'd have to build a laboratory from scratch herself.

Off to the side of the Welckers' one-story brick home was a tan tin trailer. It had originally been erected for Kelydra's grandmother, who'd fallen ill with cancer when Kelydra was four. In spite of the grim circumstances that would soon bring them into close proximity, Kelyda had looked forward to living near Grandma, who once astounded her by standing on the front porch, index finger outstretched, until a tiny hummingbird landed there of its own volition. Seeing this, Kelydra decided that when she grew up, she'd be the kind of person hummingbirds deemed worthy of landing on. The trick, Grandma said, was to move slowly, speak softly, and show respect for all living things, regardless of their size.

Kelydra's grandmother passed away before she could move in. The trailer remained vacant, steeped in bad memories the Welckers would rather avoid. But recently, the trailer had been calling to Kelydra. Perhaps it was Grandma, peering down from heaven, hummingbirds perched on her shoulders, pointing out that it was high time her trailer be used for *something* rather than going to waste. So Kelydra procured the key and got to work. After hearing that her high school science class was undergoing remodeling, she nabbed twelve boxes of old beakers, flasks, and other castoffs. When a local hospital flooded and unloaded some equipment, she got an advance on her allowance and acquired a microscope, a centrifuge, and other high-end instruments for $250.

Kelydra used the rest of the trailer's nooks and crannies for tanks, and filled them with pond water. She grew duckweed and algae, and stuck everything possible under the microscope, fascinated as she watched the green bean-shaped chloroplasts move in circles, producing oxygen. She raised hydras, creatures that resemble sea anemones but are mere millimeters high, and was once riveted for a full half hour watching one traverse the tank doing somersaults. Unlike Kelydra's bedroom, which was an explosion of clothes and sci-fi and fantasy books, from *The Lord of the Rings* to the pun-filled books by Piers Anthony, her lab was immaculate.

Day by day in her lab, Kelydra slowly began to piece the PFOA puzzle together. To protect herself from the chemical, Kelydra wore rubber gloves and goggles over her glasses, and washed her hands with distilled water afterward. Her first challenge was to figure out how to accurately gauge PFOA levels in water. This was no easy task. DuPont didn't disclose how they conducted their tests, although Kelydra had heard from the water treatment facility that it cost $5,000 per run. She couldn't fathom why it was so expensive, but it didn't bode well for her odds of replicating this test in her trailer.

Kelydra started chipping away at this question by reading every article about PFOA she could find. One fact that caught her eye was that PFOA was a surfactant, which meant it produced foam when shaken, like bubble bath. Based on plenty of bubble baths she'd had as a kid, Kelydra knew that the more bubble bath you used, the more bubbles you got. Could the solution be as simple as boiling down the PFOA-laden water, shaking it, and measuring how much foam was on top? Kelydra gave it a shot, boiling and shaking water samples for days, then weeks. After plotting her results on a graph, she found that she could determine a water sample's PFOA levels with 92 percent accuracy compared to confirmed reports. What's more, she'd essentially done it for free—a bargain compared to the alleged $5,000 typically paid per test.

"You just boil it and shake it," Kelydra explained during a studio interview on WTAP, Parkersburg's local television news station, in her effort to get the word out. Jugs started showing up on the Welckers' front porch, from neighbors concerned about the PFOA levels in their own water supplies. Kelydra also went on the air to explain how community members could limit their exposure levels. Drinking bottled water was one way, but for families who couldn't afford that, she came up with other ideas. For one, since PFOA floats to the top of water, she warned that the worst time to drink from a well was when it was running low. Well owners could also pour their drinking water into a container with a spigot at the bottom, to avoid drinking the most heavily laden PFOA water up top. By explaining the dangers of PFOA and how to avoid it, Kelydra felt she was doing a public service and helping protect her community. Her family and friends, however, didn't always hold such a rosy view of her work.

"What the hell are you thinking?" kids at school asked, coming up to her in the cafeteria. "Are you trying to turn Parkersburg into a ghost town, or what?" At DuPont headquarters, Kelydra's sister, Bonnie, endured countless coworkers coming to her cubicle throughout the day, parroting Kelydra's slogan: *You just boil it and shake it! Boil it and shake it! Boil it and shake it!* Bonnie was furious. Even though Kelydra was eleven years younger, Bonnie often felt antagonized by her little sister, who, at five-nine, wasn't so little compared to Bonnie's five-one. It was a running joke among the Welckers that Kelydra had stolen Bonnie's extra height. But Bonnie had the edge in other ways. On an IQ test they'd taken as kids, Bonnie had scored twenty points higher than Kelydra. Bonnie could solve math equations at a glance; Kelydra had to work at it. But Kelydra, ever the optimist, saw an upside to her situation with her sister. For one, since she had to work harder for things to sink in, the information she did learn stuck. Two, her heart bled for other living things who, like her, struggled to thrive. As a result, she possessed a compassionate streak that compelled her to give these underdogs a leg up. While both sisters loved squirrels, birds, and other wild animals, Kelydra tended to spoil them by feeding them treats, while Bonnie felt it was better if they learned to fend for themselves. Kelydra knew her sister was right. Still, she'd have liked it if someone had handed *her* an apple slice every once in while. So that's the way she went through life.

During high school, Bonnie was also a formidable science fair competitor, who'd swept up state and national awards for her research on the proliferation of zebra mussels, an invasive species in the Ohio River. "Here," Bonnie told Kelydra back when she was two years old, handing her a tiny tin pail. "Go count how many mussels you can find in that area of the river over there." As a result, Kelydra learned how to count to a hundred long before kindergarten. Kelydra idolized her sister and hoped she could one day follow in her footsteps and win science fairs, too.

Bonnie helped Kelydra with her science fair projects, but once those projects started involving PFOA and DuPont, she kept her distance. Kelydra was disappointed, but she could see where her sister was coming from. Bonnie was scared. Her family was scared. To avoid implicating them further, Kelydra stopped talking about her research. Since Kelydra was usually bubbling over with what she'd done that day in the lab, dinners at home became a quieter affair. Kelydra spent more and more time in her trailer, pouring her energy into the one endeavor that would right all wrongs and fix the problem: finding a way to remove PFOA from Parkersburg's water supply entirely. Kelydra knew that the current method for removing PFOA, carbon filters, was too expensive to be easily sustained. There had to be a better way.

She drew design after design, read article after article. Eventually she started circling an idea of combining small amounts of carbon with a process called electrosorption, which uses electricity to remove particles from water. Desperate for a professional chemist's input, she broke her vow of silence and discussed the idea with her dad. After perusing her design, he shook his head.

"It won't work," he said. "The electric current wouldn't be strong enough."

Normally, this would have thrown Kelydra off the scent. But lately, she had come to realize that her parents weren't gods. They made mistakes and overlooked possibilities, just like everyone else. The grooves in their minds were narrow and deep like rivers, buttressed on their banks by the fact that they had a family to support, mouths to feed, bosses to obey, and pensions to save. But Kelydra's mind was wide open. The only way she'd know for sure whether her idea would work or not would be to try it herself.

For electrodes, she grabbed the windshield wipers from her dad's old car. To increase their surface area, she attached chunks of stainless-steel wool, then sandwiched them around a thin layer of carbon. After hooking up her contraption to a six-volt battery from Wal-Mart, Kelydra stuffed it into a large syringe, poured PFOA-laden water into the top, and waited for it to pass through, drop by drop. Perhaps, if she was lucky, her electrosorption technique would have some effect. Maybe she'd get a 30 percent removal rate. That would be nice.

Once her water sample was filtered, Kelydra resorted to her old standby, the boil-and-shake. Whatever PFOA was left would create bubbles. But Kelydra was puzzled to see no bubbles at all. That would mean that her filter had removed 100 percent of the PFOA in her sample. But that would be impossible. Assuming she must have done something wrong, Kelydra filtered another water sample, boiled it, and shook it. Still no bubbles. Kelydra repeated the process again. And again. Eventually, Kelydra allowed herself to think the thought she wouldn't have dared to entertain unless the proof were staring her in the face: *Oh my god.* Then she started saying it over and over. O*hmygod-ohmygod-ohmygod-ohmygod-ohmygod-ohmygod!*

When Kelydra emerged from her trailer, she was almost glowing, her long brown hair nearly standing on end, her glasses and goggles askew. And yet at this point, Kelydra was silent. She had to keep quiet about her discovery, at least for now. Given that no one would ever believe that she, a kid, had figured out the secret to saving her town from the scourge of PFOA, it was crucial that Kelydra stake her claim before she started crowing. There was only one way to do that, and she knew what it was. Kelydra would have to apply for a patent.

Patents are a science fair kid's ticket to being seen as legit. Without a patent, science fair projects are cute. With one, they're a force to be reckoned with. Approximately one in five kids at top science fairs have applied for patents. The problem is, actually *getting* a patent is another story. Soon after Kelydra sent in her application, she was snowed under with mountains of paperwork.

At this point, most science fair kids do one of two things: (a) They hire a patent lawyer for thousands of dollars, or (b) they throw up their hands and decide that between cramming for their SATs and finishing their college applications, fighting for their patent isn't worth it. They throw in the towel.

But if Kelydra had learned anything from her dad, sitting still for hours on a riverbank, waiting for turtles, it was patience. Slow and steady won the race. So bit by bit, Kelydra started filling out forms. Sign this. Witness that. Stamp this. Check that. At the same time that Kelydra was slogging through patent paperwork, she was also preparing for science fair, which had become an ordeal of its own. Back during her freshman year, Kelydra was dismayed to find that due to lack of funding, her county would no longer be allowed to participate in the West Liberty Regional Science Fair, the only regional fair in her area. Undeterred, Kelydra formed a regional fair of her own and, within months of going door-to-door, had amassed $7,000 in funding and thirty volunteers to run the first Mid-Ohio Valley Regional Science Fair.

During her sophomore and junior years, Kelydra qualified as a finalist for the Intel International Science and Engineering Fair 2005 and 2006. There, she met some filmmakers creating a documentary called *Whiz Kids*. Knowing a good story when they saw one, the filmmakers started following Kelydra around, urging her to approach DuPont with her findings and engage the company in a showdown. But for once, Kelydra was reluctant to step up to the plate. Frankly, she felt a little intimidated at the prospect of confronting DuPont. Based on the reactions she'd gotten from friends and family so far, part of her feared that DuPont would deem her a threat—or, conversely, laugh in her face. Kelydra wasn't sure whether it would be worse to be taken seriously or to be seen as a joke. Either way, she didn't feel ready, at least not yet.

At one point when Kelydra and the film crew from *Whiz Kids* was driving by DuPont's headquarters, they pulled over. "We just want to get a picture of the DuPont sign," they explained as they set up their camera and pointed it up at the huge red oval towering above them. Within minutes, their presence was detected by DuPont's video surveillance system. Two DuPont cops were sent to investigate, followed days later by an FBI agent standing on the Welckers' front porch. Videotaping a chemical plant, the FBI agent informed the Welckers, was considered a breach of homeland security.

"Kelydra Elizabeth Welcker, we need to talk," her parents said once their daughter arrived home from school that day. Hearing that an FBI agent had come looking for her, Kelydra reacted the way any high schooler would. She was thrilled. Her? A terrorist suspect? Yeah right! But her parents weren't amused. Yet again, they impressed on their daughter that perhaps she had taken this DuPont nonsense as far as it should go. She had made enough waves. "But I'm not trying to *attack* DuPont, I'm trying to *help* them," Kelydra pointed out, first to her parents, then to her friends, and then to the residents of Parkersburg through more television spots on WTAP. Maybe if she kept saying it, her message would finally hit home.

During her senior year, after two days straight of media coverage about her research, Kelydra was stunned to hear that the Mid-Ohio Valley Regional Science Fair, the fair where she hoped to present her findings, had been canceled. When Kelydra called the fair director asking for an explanation, he said it was due to lack of funding from local companies. Typically DuPont and General Electric were the main science fair funders in the area.

Kelydra couldn't help wondering: Had DuPont pulled out, and had it been because of her? Had she created this fair from scratch only to cause its demise? To remedy the problem, Kelydra made the rounds to eighteen schools and got more than four hundred signatures from kids asking that the Mid-Ohio Valley Regional Science Fair be reinstated. But the former science fair director refused to budge.

"Let it be," he said, then hung up. Kelydra had reached a dead end. She learned that day that no matter how hard she fought, a losing battle was a losing battle. There was nothing she could do.

But there was another science fair Kelydra could enter, although getting in would be tough. It was called the Intel Science Talent Search (STS). Unlike Intel ISEF, which accepted more than fifteen hundred finalists, STS accepted only forty finalists. The competition was steep. Nonetheless, Kelydra filled out her paperwork. The day she got the call that she'd been accepted, Kelydra screamed, and cried, and couldn't wait to share her research with this prestigious panel of judges. Only once she was there, standing in front of them, would she realize what it truly feels like to be in too deep.

"What would a sixth-dimensional world look like?"

"What is dark matter made of?"

"Here." A glass of water is placed on the table. "What do you see?"

Science Talent Search is America's oldest and most prestigious science competition. Every year since 1942, first in partnership with Westinghouse and beginning in 1998 with Intel, STS selects forty high school seniors out of 130,000 applicants nationwide to fly to Washington, D.C. There, they undergo a judging process that, even to a seasoned science fair veteran like Kelydra, seems downright surreal. In the waiting area, the tension is palpable as students are escorted into one of four rooms, where they face off against a total of twelve judges. Within these rooms, the questions range from Ping-Pong on the moon to the speed of shooting stars, from interdimensional hypercubes to card tricks. Many of the questions have no answer at all and are asked merely to see how finalists think on their feet. Many stand there stumped. Some crack under pressure. As students stagger out and start whispering, each room acquires its own ominous reputation. For Kelydra, room #3 was her downfall. She emerged with tears in her eyes. As usual, the film crew for *Whiz Kids* was there, the dispassionate eye of the camera lens trained on her. Kelydra tried to hold it together, but failed.

"What would have happened if my mom was here is she would tell me to stop crying," Kelydra said to the camera, crying. When asked what she thought her chances of winning were, Kelydra didn't hesitate. "Zilch to none," she said.

Kelydra was right. She didn't win an award at Intel STS. While she knew that just being picked as one of forty finalists was an honor, it was still a humbling experience. She had always dreamed of going to Harvard, but now she wasn't so sure. Maybe she wouldn't fit in at a big school. She was a small town gal, a DuPont brat who had bitten off more than she could chew. Maybe in the grand scheme of things, what she'd done wasn't such a big deal. Maybe she'd alienated her friends and family for nothing. For the first time in a long time, she retreated into her shell.

Then Kelydra got a call from her sister, Bonnie. "How about we go see that new *Transformers* movie?" Bonnie asked. It was a small gesture, but Kelydra knew her sister well. Given that they pretty much hadn't spoken in the two years since Kelydra had started her research on PFOA, this gesture, while small, said it all. After that, the two sisters started hanging out more often, catching up on all the milestones they'd missed in each other's lives. By this point, Bonnie was married and pregnant with twins. She still worked at DuPont, but both sisters knew better than to delve too deeply into that subject. Occasionally when someone called Bonnie, an uncomfortable silence would descend when she said she was hanging out with her little sister. You mean the one they just saw on television? Talking about DuPont and PFOA?

Eventually Bonnie confided in Kelydra that she thought her research really could help DuPont, but that her timing wasn't ideal. Maybe things with the environment needed to get worse before they got better. Or maybe Kelydra needed to grow up and go to college then revisit the issue. Kelydra, of course, was gung ho about going to college. Paying for it, though, would be tough. Through science fairs, Kelydra had scrounged together some money, but it was only enough to pay for one semester.

Kelydra did have one more science fair to go: the Junior Science & Humanities Symposium (JSHS). The fair was held at West Virginia Wesleyan College in Buckhannon, a college that Kelydra had always loved but considered financially out of her reach, given that tuition cost nearly $30,000 per year. During the fair, Kelydra struck up a conversation with a petite woman in her fifties. Kelydra had no idea who she was, but she was certain this woman wasn't a judge. Most science fair kids bend over backward to suck up to a judge, but tend to loathe explaining their work to the lowly layperson. But Kelydra didn't care whom she was talking to. She was happy to break down her research so that even a four-year-old could get it. *You just boil it and shake it.* Science was simple. It didn't need to be intimidating. After making small talk with Kelydra for a few minutes, the petite woman drifted off. At which point, a fellow competitor came up to Kelydra, slack-jawed, as if Kelydra had just done something big.

"Do you have any idea who you were just talking to?" the competitor asked. When Kelydra said no, the student informed her that she'd been chatting up the president of West Virginia Wesleyan College. *The* president? As Kelydra's mind raced through every word she'd said, it struck her that for a college president, this woman seemed so modest, so unassuming. But if there was one thing Kelydra had learned, it was that the very best gifts in life arrived in deceptively inconsequential packages.

At the awards ceremony for JSHS, Kelydra won first place. The president of West Virginia Wesleyan also took Kelydra under her wing, pointing her to scholarships that Kelydra had never known existed. Thanks to the president's help, Kelydra landed enough money to attend West Virginia Wesleyan the following year as a college freshman. Along the way, she picked up many smaller awards that inspired plenty of chuckles, like a year's supply of toilet paper from Angel Soft (the Welckers managed to make it last for two years). Or the BRICK award, which earned her a spot on the back of a Doritos bag showing her photo and a synopsis of her research.

At first, Kelydra was skeptical that the back of a Doritos bag was the best venue for sparking a scientific discussion about PFOA, especially since she'd been picked for the flavor Blazin' Buffalo and Ranch, which was so spicy people rarely bought it twice. Yet once the Doritos bags were out, emails flowed in from across the country, saying, *Hey, you're on the back of my Doritos bag! Love your research! Keep it up!* The fact that someone actually bothered to read the back of a Doritos bag *and* then took the time to email her was a revelation to Kelydra.

But Kelydra's biggest coup was yet to come.

. . .

Kelydra was sitting on her front porch, staring down her hundred-foot gravel driveway, following a tried-and-true Welcker tradition: She was waiting. After two years, mountains of paperwork, and more than a few weak moments where she considered calling it quits, she was surprised to learn that her patent for removing PFOA from water had been approved and would soon arrive in the mail. Every afternoon, Kelydra waited on her front porch for the postal delivery lady, who swung by at three P.M. Occasionally, the family cat kept her company, asking for a scratch on the chin, or rolling onto her side for a belly rub, reminding Kelydra that there was nothing simpler than a cat's life. Cats got fed and sat on laps. Nothing bad happened to cats, provided they had humans to take care of them. *Lucky bastards,* she thought.

But today was also Kelydra's lucky day. In the distance, she could see that the mail lady wasn't just dropping off the mail in the box like she usually did, but was striding up their long, gravel driveway carrying an official-looking envelope, which read: *Do not bend.* "I knew you were waiting for his," the mail lady said with a smile. Kelydra was so thrilled she couldn't speak. All she could do was squeak.

Inside the envelope was a five-page booklet with a ribbon around it, stamped with an official patent emblem. Inside the booklet was a synopsis of the instrument she'd created that could remove PFOA from water. The device still needed work— for one, it was slow (it took an hour to run one glass of water)— but it was a start. Success doesn't happen overnight.

"Mom! Mom! Mom! I got it!" Kelydra hollered, running into the house. Her mom, Karen, hearing the news, started jumping up and down, too. "What's all the hubbub about?" her dad, Pete, asked, trudging upstairs from his woodworking shop in the basement, where he carved duck decoys out of old pieces of driftwood. Once he heard about the patent, he smiled.

"Well!" Pete said. "Congratulations, kid. You deserve it." Even though he wasn't overflowing with enthusiasm, his small nod of approval meant everything to Kelydra. It had been a long time since her dad had discouraged her from pursuing her work on PFOA. That's because in the same way that Kelydra had learned patience from her dad, he had learned something from his daughter as well: The DuPont Rules didn't need to be followed to a tee. Sometimes, you do the assignment that is given to you. But other times, you ask questions.

For now, Kelydra is done standing up to DuPont. After graduating high school in 2007, she hung up her boxing gloves and hunkered down to learn, and grow. That same year, DuPont announced their success in reducing PFOA emissions in the U.S. by 98 percent, and a commitment to eliminate the need to make, buy, or use PFOA by 2015. In addition to changing the carbon filters regularly at municipal water plants, the company installed filters in residents' private wells or offered them bottled water so they, too, could remain safe. Due to the EPA's efforts to crack down on PFOA, the U.S. Centers for Disease Control and Prevention reported that human blood levels of PFOA collected in 2003–2004 showed a 25 percent drop from levels found in samples collected in 1999–2000. Expectations are high that this trend will continue.

By the time I heard her story, Kelydra was a sophomore at West Virginia Wesleyan in Buckhannon, studying chemistry. One day, she hopes to improve on her patent and approach

DuPont with her ideas so that she can eliminate PFOA from Parkersburg's water supply for good.

"What I learned was that science isn't just someone standing up in a white lab coat, giving terms ten syllables long," Kelydra said. "To me, what science is, is you take that research and give it to people who really need it."

And not just people. Kelydra wants to preserve the rivers and lakes she's been wading into for as long as she can remember, and the turtles, toads, snakes, salamanders, and other wildlife that swim in those waters. Because every living thing, no matter how big or small, is equally important. Whenever Kelydra feels like she's lost sight of this pearl of wisdom passed to her by her grandmother, she spots a hummingbird hovering outside her front porch and remembers. Has she followed in her grandmother's footsteps and convinced a hummingbird to land on her finger?

"Not yet," Kelydra laughed. "My sister almost did. Then she sneezed."

But knowing Kelydra, I have no doubt she will keep trying.

7

THE MELTDOWN

Hearts will never be practical until they are unbreakable.
—THE WIZARD OF OZ

It was nine A.M., time for science fair coach Tanya Vickers to do a final head count on her all-star team. Only at AMES—short for the Academy for Math, Engineering & Science—kids don't play your traditional sports. Head into the bathrooms at this small Salt Lake City public school, and the graffiti scrawled on green-tiled walls says it all: *It smells like CH_4 in here!* (CH_4 is the chemical makeup of methane.) Or: *I pity the fool who doesn't know MySAL!* (MySAL is a computer programming language.) One year, students from AMES spray-painted a parking lot with a cryptic message: *f/m f/(dv/dt) hr QT > west.* That's science-ese for a dis on their rival school, West High.

At AMES, science fairs are the sport of choice. Which is why Tanya was surprised to see that one of her star players, Sarah Niitsuma, was absent from class. Sarah's teammate, Shwan Javdan, volunteered to go find her. As he made his way down to the school's West Pod, a lounge area filled with computers, he spotted her instantly. Sarah was easy to spot. She was strikingly pretty, with traces of Mexican and Japanese in her features, and dressed to turn heads in cute skirts and frilly tops, her curly brown hair flowing down to her waist. But Sarah's coup de grâce was her tiara. She wore tiaras to school on such a regular basis that the high school principal, Dr. Church, typically greeted her with "Hello, Princess" when they passed in the hall. Sarah always looked poised, perfect, put together. But Shwan was about to see a different side of Sarah.

"What's up?" Shwan said, sitting down next to her. "Why aren't you in class?"

Sarah ignored him. She was busy rewriting Shwan's science fair write-up word for word. She had been working on their project all night, alone, fueled by Monster energy drinks and surges of panic and indignation. Sarah always knew that doing a team project was a bad idea, that Shwan would drop the ball. Now she was paying for it.

"Your write-up is terrible," Sarah finally informed Shwan after he continued pressing her for an answer. "We're never going to win. I don't want to do this. I give up. It's over."

"It's not *that* bad," Shwan said, shocked by Sarah's dramatic take on the situation, but by now Sarah was blinking back tears, so Shwan backed off, and reported to Tanya, who winced when she heard what was up. In a mere three hours, her students would need to head down to SkunkWorks, a room named after a highly secretive research facility at aeronautics firm Lockheed Martin, to present their research. Everyone's projects were ready to roll—except Shwan and Sarah's. Their tri-panels were blank, barring a few stars and a Kermit the Frog sticker Sarah had affixed to the edges. Unless by some miracle they could assemble their boards in record time, their chances of winning, or even competing, were next to nil. Still, there was time. There was hope.

"Let's go get her," Tanya said to Shwan. "We've got work to do."

· · ·

On the brink of the big event, it's inevitable: Science fair competitors can implode under the pressure. The heavy workload, combined with the specter of scholarships just out of reach, whips certain kids into such a frenzy that they pop. During my sojourn at the Intel International Science and Engineering Fair 2009, I heard plenty of tragic tales of kids who cracked before crossing the finish line. Sarah and Shwan could easily have ended up among these science fair casualties. But in their case, teamwork prevailed.

For Sarah, the vectors leading up to her meltdown in the West Pod on the day of the AMES science fair were set in motion at age seven, when her dad opened a savings account in her name. "This is for when you go to college," he said. *College.* Sarah had never heard the word before, but it sounded important. Sarah's parents had started college, but neither of them had finished. Her dad, Donovan, worked at the post office, her mom, Rebecca, as a nurse. Sarah knew, without it having to be said, that her parents wanted more for her. In fourth grade, she came home from school to find that those plans had changed.

Sarah entered her bedroom to find it completely empty, as if she'd never existed. "Your mom and I are getting a divorce" was Donovan's abrupt reply when Sarah asked him, terrified, what was happening. Before Sarah or her two younger siblings could protest much, he packed them in the car, explaining that he needed to rent out their rooms to make money, and dropped them off at the Niitsuma Living Center. This was where Rebecca worked, caring for a handful of occupants. Now it was also where she'd raise her family. Unable to rent a place on her own, Rebecca had managed to squeeze two small beds into her office. Sarah and her sister Danielle slept in one, Rebecca and her son Donny in the other.

Up until this point, Sarah had been shielded from the grimmer aspects of her mother's profession. Now she was drowning in it. Rebecca, dressed in scrubs, her hair up in a bun, was perpetually in motion tending to her patients, who were elderly, bedridden, and needed help with the simplest of tasks, like cooking and going to the bathroom. Sarah was shocked to discover that there were diapers for adults. At first, she tried to make the best of it and bond with her new family. Every day, she'd spend some time with an Alzheimer's patient named May, who'd sing songs and hand her a stuffed animal, which Sarah would return later so that May could give it to her again. When May died, Sarah learned the hard way that there were risks in getting too close. From then on, she kept her distance.

Sarah also struggled to adjust to her tight quarters. Since all four of her family members lived in one room, privacy was not an option. For Christmas, Rebecca got Sarah a fold-up room divider so she could cordon off a small area, but this wasn't enough. Sarah started doing her homework in the bathroom, sitting in the shower. Since this was the same bathroom used by patients, she started doing her homework in the middle of the night. During the day, she often slept, tired or depressed, or snapped at her siblings. They all got on each other's nerves, like too many zoo animals trapped in too small a space.

Donovan phoned every few weeks, but Sarah only grudgingly took his calls or came by to visit. Since she'd been kicked out, her idolatry of him had faded. With distance, many of his actions even struck her as mildly sadistic. Once while they were visiting an ostrich farm, he had relished dangling her over the fence as she shrieked in terror and ostriches nearly nipped at her feet. After learning that Sarah was afraid of grasshoppers, he caught the insects and threw them on her just for kicks. When

Sarah scraped together the money to buy him an Old Navy shirt in seventh grade, he rarely wore it, preferring instead to wear T-shirts with holes in them, tucked into pajama pants over his gut. Once he told Sarah she would have ended up prettier and smarter if she'd had a different mom. When Sarah asked him why he and her mom divorced, he replied, "She thinks you're God's gift to the world and I don't." It was her dad's form of a joke, so Sarah laughed.

Amid the chaos her life had become, Sarah did have one escape. She could walk three miles to her older sister Raquel's house and crash on her couch. Raquel had a daughter named Rachel, who was technically Sarah's niece. But since they were the same age, they'd been best friends for as long as Sarah could remember. They went to the mall and bought matching outfits. They often studied together. Back in elementary school, Sarah struggled with reading to the point that teachers considered holding her back, but Rachel, who could read with ease, drilled Sarah every day until she caught up.

Raquel lived with her boyfriend, Tony, who worked in construction. Unlike Sarah's dad, who was scruffy and unkempt, Tony dressed meticulously and kept his face and head clean-shaven. For Christmas, when Sarah got him some knockoff Calvin Klein cologne, she could smell it on him every time she was over. Tony brimmed with gifts like Barbie dolls and told Sarah that she was pretty, a compliment that never struck her as unusual. Then one night in eighth grade, when Sarah fell asleep on the couch watching TV, she half woke up to sense Tony sitting beside her.

As his hands wandered under the covers, Sarah shrank away from his advances, but his hands followed. Sarah couldn't bring herself to open her eyes, but she could smell her Calvin

Klein cologne and hear the comforting patter of a cartoon on TV. Unsure how he would react if he knew she was awake, she pretended to be asleep, tossing and turning, until he left her alone.

The next morning, as Rachel made breakfast, Tony smiled his usual smile at Sarah and asked, "What were you dreaming about last night? You were really thrashing around." Startled, Sarah didn't know what to say. Maybe Tony was right. Maybe it was just a dream. When Sarah told Rachel what had happened, she, too, suspected it was just a dream. Sarah hoped they were right. After all, Raquel and Rachel loved and depended on Tony. Without him, they'd be devastated. But as Sarah continued sleeping over, Tony came to the couch again and again. By the fifth time, he managed to undo her pants, shimmy them down to her knees, and lie on top of her before she managed to push him off, still pretending to sleep. That's when Sarah realized she had to tell, even if it splintered her already piecemeal family further.

One night, while watching TV with her big sister, Raquel, in her bedroom, Sarah waited for the right moment. Only after Raquel had closed her eyes and drifted off could Sarah say the necessary words: "Tony's touching me." Sarah half hoped Raquel wouldn't hear her, but as her sister's eyes flew open and filled with tears, it was clear that she had. As she cried, questions came tumbling out of her. What did Tony do? How long had this been happening? What should they do now? Both agreed that they shouldn't confront Tony until Sarah had talked to the cops, which Sarah did the next day, approaching a guard at school. For the next few weeks, Sarah was whisked away to an array of appointments with doctors and detectives, who trained a video camera on her and peppered her with uncomfortable questions. *Where did he touch you? How did it feel?* Afterward they handed her a stuffed animal, a brown dog, "for being brave." Sarah did feel brave. But she also felt scared. What if she'd said something wrong?

Her mother, Rebecca, cried and tried to hug her daughter, but Sarah pushed her away. She wasn't in the mood to be smothered with affection. Weeks earlier, Rebecca had asked her daughter, "Is someone older trying to do something to you?" She had a sixth sense about her daughter, which often rattled Sarah when it hit too close to home. Sarah said no, everything was fine. Even now, she desperately wished that everything would return to normal. But it was too late for that. Sarah had opened her mouth, and now everyone in her family would have to suffer the consequences.

Tony was arrested, convicted of attempted aggravated sexual abuse of a child, and sent to prison. At his court hearing, Sarah didn't take the stand, but she did write a letter, asking the judge not to put him away for a very long time. He'd made some mistakes, she wrote, but had taken good care of his family. Sitting in the courtroom pews, Sarah saw Tony for the last time. He glanced her way only once, for a second. In Sarah's mind, he looked furious, as if she was the one who'd done something wrong. Maybe she had. This sliver of doubt only grew once she got a phone call from her dad.

He fumed. "What the hell were you thinking? Why did you let Tony do that?"

"I didn't!" Sarah wailed. She had tried to fight. But why didn't she fight harder? Why didn't she just tell Tony to stop rather than tattling to the police? Even her best friend Rachel seemed skeptical.

"Why didn't you just say no?" Rachel asked. Sarah could hear the subtext loud and clear: *Why did you ruin our lives?* As time passed, Rachel's questions gained more of an edge. Sarah stopped visiting her so often, at which point Rachel started texting her. *Why did you let him do that? You must have liked it.* Eventually, her comments cut so deep and came so relentlessly that Sarah changed her phone number and avoided Rachel entirely.

Even so, Sarah missed her old best friend. Once, Sarah called Rachel just to hear her say "Hello?" before Sarah hung up. As years passed, she heard that Rachel dropped out of school and moved in with her boyfriend. The two girls started bumping into each other at family events, and while things were cordial, they never bonded like they had as kids. And besides, these days, Sarah found that she had less and less in common with Rachel anyway. Rachel was a high school dropout. Sarah took school seriously. Good grades meant she'd go to college. College meant she could get a good job, and her own place. She could sleep in her own bed. She wouldn't need anybody. Her life had been one long lesson in how the people you depend on will inevitably let you down. As a result, Sarah decided to remain aloof and untouchable.

While wandering around the mall alone one day, Sarah spotted a tiara at a jewelry store. Back at her dad's house, her mom had once hung a tiny plaque on the wall that said *Sarah: Princess,* since Sarah meant princess in Hebrew. Sarah had no idea where this plaque was now—perhaps her dad had trashed it—but that didn't mean she couldn't be a princess on her own terms. She bought the tiara and started wearing it to school. She bought more tiaras, and had been wearing them ever since. Aside from her name, Sarah wasn't sure why she liked tiaras so much. Perhaps it was because they made her feel special. Individual. Above it all.

Sarah knew, based on finding her dad's bank statements, that the college savings account he had opened for her when she was seven contained no more than a couple hundred bucks. In science class, Tanya had informed her students that science fair winners often won scholarships. *Scholarships.* Science fairs could be Sarah's way out. Then Sarah's science teacher Tanya Vickers threw Sarah a curveball that would complicate things.

With four years as a science fair coach under her belt, Tanya took pride in her ability to defuse a meltdown in the making. This required careful surveillance, so Tanya went all out to put her students at ease. She told them to call her Tanya rather than Mrs. Vickers, to convey that they were on even footing. She invited them to call or email evenings and weekends, disrupting dinners and nights spent with her husband and two kids. Tanya was also wary of letting kids do team projects, since arguments inevitably erupted over who was doing more work. Yet in certain circumstances, team projects could benefit all parties involved, and Sarah's situation seemed perfect for one.

Sarah had started off working solo on her project, which examined microbe levels on infant changing tables in public restrooms. It was an ambitious project, and she had fallen behind in her work. It didn't help that, in an effort to scrounge together some money for college, she had started working as a tagger at an army-navy surplus store, as a math tutor at school, as a Sunday school teacher at church, and at any other odd job she could pick up.

Tanya knew that Sarah desperately needed a scholarship. She also knew that the competition among the individual projects was stiff that year, so odds were slim that Sarah would stand out. Team projects, however, were scarce, which meant the awards in the team category were more easily up for grabs. If Sarah truly wanted a scholarship, she'd be better off teaming up with a partner, and Tanya knew the perfect person for the job. Shwan Javdan.

After hearing Tanya's suggestion, Sarah cringed. For one, Shwan was a sophomore, Sarah a senior. Two, she just wasn't comfortable teaming up with anyone. It didn't help that ever since the Tony incident four years earlier, Sarah had avoided boys. If one sat too close to her at lunch, she slowly edged her seat away from him. The thought of being touched by a boy made her feel sick. Whenever she caught a whiff of Calvin Klein cologne, she wanted to scream. Still, if she wanted to do well at science fair, Sarah would have to step off her pedestal, swallow her pride, and join forces with Shwan. Tanya had emphasized that it was Sarah's choice, but the reality was that she had no choice. If she wanted to win, she knew what she had to do.

For a while, Sarah and Shwan worked side by side silently, divvying up the duties. Unlike most guys in class, who often tried to flirt with Sarah, Shwan made no such attempts, which suited Sarah just fine. Sarah was in charge of swabbing samples at public restrooms and testing them at the lab; Shwan was in charge of writing up the results. As Sarah rode the bus throughout Salt Lake City and the surrounding suburbs to conduct her research, occasionally she felt like she was doing more work than Shwan. But rather than pointing this out, she kept quiet, and let her resentment simmer. Speaking up, in her experience, was only a recipe for trouble.

On the night before the AMES science fair, Sarah asked to read what Shwan had written. He was a very good writer, so Sarah had trusted he'd hold up his end of the bargain. Only once she read it, it quickly dawned on her that her trust had yet again been misplaced. His write-up was good, but she needed it to be great. As Sarah read sentence after sentence, she saw the scholarship she'd dreamed of winning slipping through her fingers.

Enraged, Sarah grabbed a red pen and started marking up Shwan's write-up. She stayed up all night making changes and putting together their board, desperate to get back within reach the scholarship she needed so badly. Sarah went to school the next morning, skipped Tanya's class, and sat slumped over a computer in the West Pod, the text from Shwan's write-up swimming in front of her. Sarah had had it. She hated Shwan, and science fairs, and the cards that life had dealt her. Winning science fair was her ticket to college, and college was her ticket out of her miserable life. And today, she would lose it all.

The instant Tanya saw the look in Sarah's eyes that morning, she knew that unless she trod carefully, Sarah's meltdown would reach epic proportions. "Let's go down to Dr. Church's office," Tanya suggested softly to Sarah and Shwan, who was hovering nervously behind them. If Tanya was going to salvage their situation in time for the science fair in a few hours, she'd need to take them to the one person who could solve the problem fast. The high school principal.

Within his office, Al Church had a fried egg paperweight on his desk. It was there to remind him that no matter how hard he tried, he'd always end up with egg on his face, that things were bound to get messy before they got better. As a former food critic, Dr. Church abhorred fast food, a belief he'd carried over into his educational philosophy as well. Huge high schools, in his opinion, churned out lackluster students much like McDonald's spit out Big Macs. Here at AMES, where he'd served as principal since the school's inception in 2003, Dr. Church had hoped to do something different.

AMES was one of more than two hundred "early college high schools" across the nation funded by the Bill and Melinda Gates

Foundation, offering college courses to help kids get a jump on their college education. Capped at five hundred students, this small school was geared toward serving traditionally underserved populations, such as ethnic minorities and low-income families. As Dr. Church sat down with Sarah, Shwan, and Tanya, this personalized approach to schooling was about to work its magic. In this particular instance, Dr. Church likened it to marriage counseling.

"I've been to marriage counseling, and I think it's very helpful," Dr. Church volunteered, as if to show that even he, as the principal, wasn't perfect. He turned to Tanya. "How about you? Been to marriage counseling?" Tanya smiled and said yes. "The thing that's great about marriage counseling is that it'll get you two talking without attacking each other. Now, I'm going to ask each of you to explain your side of the story, and you can't interrupt. Sarah, you go first."

Sarah had been to counseling once before. It was back in eighth grade, after the Tony incident. At first, all Sarah could manage to say to the therapist was that she felt fine. Occasionally, she walked out in the middle of her sessions, saying her dad thought counseling was stupid and a waste of time. But as her therapist indicated that her counseling sessions would soon come to an end, Sarah knew that if she didn't speak up, the window of opportunity she had to be heard would close for good. So Sarah talked about what Tony had done to her and how it made her feel. She unloaded the guilt she felt for ruining Raquel's and Rachel's lives, which made her regret telling on Tony at all. She opened up about her dad, wondering why he didn't care about her. As Sarah talked, she could magically feel a weight lift. She continued therapy, twice a week, for more than two years.

Now it was time for Sarah to speak up again, so she did. She talked about how hard she'd worked on science fair and how, in her eyes, Shwan wasn't pulling his weight. Next it was Shwan's turn to talk. The first thing he said was that he was sorry, that he had had no idea that Sarah felt the way she did, and that if he had, he would have gladly done more. Sarah knew Shwan meant what he said. Shwan wasn't a bad science fair partner by any means. He was smart and organized and reliable, provided Sarah could convey her needs.

After they'd both said their piece, Shwan and Sarah sat there, shocked at how quickly the tension had been diffused in the room. "Thank you both for coming here," Dr. Church said, smiling. "Now, why don't you two go finish your science fair project?"

For the next two hours, Sarah and Shwan printed out graphs and charts and pasted them onto their board. Other kids in class pitched in, so that by the time science fair rolled around that afternoon, their board had miraculously squeaked into Skunk-Works just in time. Before the judges started circling the booths, Tanya drilled her students one last time. Tuck in your shirt. Turn off your cell. When in doubt, call the judges *Doctor* So-and-so. With judges, it was always a good idea to kowtow a little. Every little edge counts.

During their presentation, since Sarah and Shwan hadn't had time to practice who would say what, Shwan rattled on to the point that Sarah felt sidelined. Sarah piped up after one judge had drifted off. "That's *my* part—I'm supposed to say that." Their relationship might have improved somewhat since their therapy session four hours earlier, but they were still pretty dysfunctional and working out the kinks. Which made it all the more surprising when, at the awards ceremony, Sarah and Shwan were announced winners of an Outstanding Achievement Award and the Most Enthusiastic Award. *Is that a joke?* Shwan wondered, finally concluding that they must have put on a more convincing act of being happy together than he'd thought. Whatever the reason the judges had for giving them these awards, all that mattered was that it was enough. By doing well at the AMES fair, they'd managed to advance to the regional Salt Lake Valley Science and Engineering Fair, which was coming up in a few weeks. If they did well there, they had a shot of qualifying for the Intel International Science and Engineering Fair 2009.

As Sarah's and Shwan's names were called for their awards, Sarah should have been thrilled. But at that moment, she was slumped in her chair, asleep, and had to be shaken awake to join Shwan on stage. Days later, as Sarah's lethargy worsened, doctors diagnosed her with mononucleosis and ordered her to bed. Sarah slept for a week straight. When she woke, she found out that the extra lab work they'd needed to do before the regional fair was done. Shwan had picked up the slack.

"Thank you for doing that," Sarah said to Shwan once she'd arrived back to class.

"It was no big deal," Shwan said. But to Sarah, it was a big deal. She wasn't used to people in her life coming through for her. She began warming up to Shwan and offered to come by his house so they could polish their presentation. Unlike her own home, where Sarah could never find a moment's peace, Shwan's house was pin-drop quiet and sparkling clean. Shwan's mother, who worked in accounting and finances at the University of Utah, greeted Sarah warmly. Sarah was shocked and envious to discover that Shwan, an only child, had his own bedroom *and* his own study. As they worked, Sarah looked around the room, trying to get a sense of what Shwan's life was like.

"What are those?" Sarah asked, pointing to a line of different-colored belts on the wall. Shwan explained that they were his martial arts belts. Shwan had a black belt in tae kwon do and was halfway toward his black belt in aikido. Sarah pointed to a map of Iran on the wall, asking if that was where Shwan's parents were from. Shwan said yes, adding that his parents had left Iran for many reasons. At that time, Iran's government was slowly turning into a theocracy. Muslims, the main religious group in Iran, were growing more agitated and extreme, while Shwan's family came from a pacifist religion known as Baha'i. Most importantly, the educational opportunities were much better in the United States, a fact that Shwan had grown up knowing he should never take for granted. To this day, Shwan has consistently kept a 4.0 GPA.

As they worked, Sarah waxed on about how she was absolutely certain they would win at the regional fair. Shwan pointed out that it was possible, but that they should remain realistic. Sarah knew he was right, but she couldn't help hoping they would win—and that her dad might be there to see it.

In the past few years, Sarah's relationship with her dad, Donovan, had improved. Since her mom's home office setup at the Niitsuma Living Center had become too hectic for Sarah to concentrate on her schoolwork, and going to Raquel and Rachel's was no longer an option, Sarah had tentatively asked Donovan if she could sleep on his living room couch, and he had acquiesced. His home was quieter but had its own set of problems. For one, the house had fallen into complete disrepair. Many of the doors no longer had doorknobs. The living room was half-painted, but had never been finished. Stacks of dirty dishes filled the sink, but Donovan refused to let anybody load the dishwasher, since no one ever did it quite right, which would compel him to angrily yank every dish out and put it back in, just the way he liked. He also owned two dachshund mixes, Amanda and Chloe, but didn't bother to let them out that often. As a result, urine stains covered the carpet.

Sarah had recently decided to get a dog of her own. Isabelle, a black-and-white terrier, was just a puppy when Sarah got her, and she followed Sarah everywhere. If Sarah and her dad went boating, Isabelle would stand at the shore and bark, and once she jumped in and came swimming after them. Sarah and Isabelle were inseparable. Then one day, Sarah came home after work to find that Isabelle wasn't there waiting for her.

"Where's Isabelle?" she asked her dad, who shrugged, then sequestered himself in his room, where he spent most of his time trolling Internet dating sites. That night, Sarah walked the empty streets in her town, yelling Isabelle's name, before returning home empty-handed. Something wasn't right. Her dad wasn't telling her the whole story. The next day, Sarah picked up his cell phone when he wasn't looking and scrolled through the list of outgoing calls. Spotting a number she didn't recognize, Sarah called it. A man answered. Sarah introduced herself, then asked the question that had been bugging her all night. By chance, had he just purchased a black-and-white terrier puppy?

After a moment's pause, the man said yes. Sarah's dad had sold her dog to a stranger for fifty bucks. The unnecessary cruelty of it hit her like a kick in the stomach. Just as she was about to beg to buy Isabelle back, the man started rambling on about how he lived out on a sheep farm in the country and that Isabelle seemed happy. Sarah knew that if she brought Isabelle back there, her dad would most likely sell her again. Sarah could move back in with her mother, but given the chaos there, her schoolwork would suffer. As much as it broke her heart, she had to admit that Isabelle was probably better off where she was.

Sarah never saw Isabelle again. Sometimes she flips through old photos of her puppy or holds Isabelle's old blanket up to her face, breathing deep. When Sarah confronted her dad, he claimed that Isabelle wasn't house-trained anyway, so if she was gone, good riddance. Sarah would never forgive her dad for what he did. And yet, even then, she held on to the hope that one day he might pleasantly surprise her. So Sarah invited her dad to the science fair by pinning the invitation on his fridge, where she knew he would see it. Donovan thought science fairs were a waste of time, but Sarah wanted to show him otherwise.

At the awards ceremony at the regional science fair, Sarah scanned the audience for her dad but didn't see him. Maybe she should have asked him to come, rather than relying on him spotting the invitation on the fridge. Sarah was disappointed, but not surprised or as sad as she'd thought she might be. That's because she knew there were people in her life she could count on. Her mom, for one, was sitting in the audience beaming. So was her science teacher, Tanya Vickers. Last but not least, Shwan Javdan had stood by Sarah, steady as a rock, and in spite of her meltdown would end up writing in her yearbook, *You are the best science fair partner ever,* in his impeccable cursive script.

That day, at the regional fair, Sarah and Shwan won Grand Prize. That meant they would soon hop on a bus to Reno, Nevada, and present their work at the Intel International Science and Engineering Fair 2009, where they would have a shot at being showered with scholarships. For most of her life, Sarah had learned that there were risks entailed in getting too close to people. But that day, she learned that there were people in her life who were worth letting in. She didn't need to be a princess in a tiara, above everything and everyone, at least not all the time.

Sarah turned to Shwan and hugged him. The thought of any physical contact with a boy still made her a little uncomfortable.

But Shwan was different. She couldn't have made it up on stage without him.

On May 10, 2009, Sarah and Shwan boarded the bus to Reno. They were heading to the pinnacle of science fairs and had made it there against all odds. "The team dynamic always tends to be caustic," said Tanya with a sigh, adding that Dr. Church's counseling had worked wonders. "Sarah needed to unload. Coming from her circumstances, this was empowering. She learned that speaking up, although it's difficult, really could have changed the course of things."

During their eight-hour bus ride to Reno, Sarah and Shwan talked about their lives, although he would never hear her full story. Few people would ever know everything Sarah had been through; there was just too much to tell. For Sarah, it was enough to know that Shwan was there for her if she needed him. "I'm glad I had a team member," Sarah told me. "If he hadn't done that extra work, our project wouldn't have been as good."

But was their project good enough to win at Intel ISEF? Would Sarah's dreams of receiving a college scholarship come true? Sarah wasn't sure. But the one thing she did know was that she and Shwan had made it this far together. And she was looking forward to seeing where they'd go next.

GLOVE BOY

First they ignore you, then they laugh at you,
then they fight you, then you win.

—MOHANDAS K. GANDHI

On March 28, 2002, all sixteen hundred students at Central High School in Grand Junction, Colorado, were summoned for a special assembly. Why was a mystery, but the one thing they did know was that whatever the assembly was for, it was big. Security had shut down the parking lot so no one could slip out early. The auditorium stage overflowed with flowers. The cheerleaders were in uniform. The band was poised to play. On the floor, a red carpet had been rolled out, as if George Clooney were about to stroll in and accept his Oscar with a smile.

Given that the assembly was put together by the principal and former basketball coach Denny Dearden, kids whispered that maybe this big to-do was a pep rally for the baseball team, which had recently won the state championship. Denny, however, had decided to applaud a new kind of hero at Central High. He wasn't sure how the students would react. But he was willing to roll the dice.

On cue, the band started playing. The cheerleaders started kicking. The doors to the assembly hall opened, and down the red carpet came Ryan Patterson. This unassuming eighteen-year-old with a beanpole build wasn't an athlete—far from it. Instead, he'd made a killing on the science fair circuit, earning more than $420,000 in cash and scholarships for inventing a glove that could do something extraordinary. Most students had heard about Glove Boy, and as Ryan took center stage, students leapt to their feet and started cheering.

"RYAN ROCKS! RYAN ROCKS! RYAN ROCKS!"

Days later, Ryan's trophies filled their own display case in Central High's hallway, next to trophy cases for the football team, baseball team, and other sports. Ryan had proved that science was also a sport and that winning could bring as much fame and glory as what is showered on an all-star athlete. Kids passing by this shrine to Glove Boy stared through the glass and wondered: Who *is* Ryan Patterson? How did he single-handedly turn the uncool pursuit of science into a red carpet–worthy, trophy-winning event? I headed to the small, unassuming town of Grand Junction for answers. And what I found was light-years from what I'd expected.

Ryan first started down his strange, circuitous path to science fair stardom at age one, crawling toward a power outlet, a butter knife clamped in his chubby hand. The knife was too dull for Ryan to do any damage to himself with it alone, but it was perfect for the purpose he had in mind for it. Plopping down on his diapered bottom, Ryan pried off the wall socket's plastic protector. At this point, he was face-to-face with the two empty eyeholes and the slotlike mouth of the power outlet itself. Where did those holes go? To find out, Ryan inserted the butter knife and started digging.

His mom, Sherry, was in the next room preparing dinner when she heard an electric *pop,* then turned to see a butter knife flying through the air, glinting from a bright burst of light. Sherry dropped the vegetables she'd been chopping and ran toward Ryan. The first thing she saw, turning the corner, was that the wall, once white, was now scorched black. Then she saw Ryan sitting there, screaming. Sherry picked him up, relieved to see that he had survived the shock, praying that this experience would dampen his interest in electricity. No such luck. At age two, Ryan was seated on Santa's lap. When Santa asked the blond, blue-eyed boy what he'd like for Christmas—a Tonka truck? A Play-Doh Fun Factory, perhaps?—Santa heard his first real head-scratcher of the season.

"Extension cord."

Ryan's parents shrugged, then tried to look on the bright side. Their son might be strange, but at least he was easy to please. Come Christmas, there it was, wrapped up under the tree, along with a lightbulb, another item on Ryan's wish list. For the next year, wherever Ryan went, an eight-foot-long extension cord traveled with him, neatly coiled—to the grocery store, to Grandma's, even to bed. At age three, Ryan asked Santa if he could upgrade to a multi-plug with a light-up switch. At age four, he asked Santa for a chain saw. A real one, of course. That's where the Pattersons drew the line, but it didn't help much, since at this point, Ryan was old enough to reach for whatever caught his eye on tables and countertops. His first casualty was the electric can opener.

Sherry entered Ryan's room one day to see her electric can opener pried in half, its wires, gears, and springs strewn on the floor. "Ryan," she scolded, "you can't just take my things apart." Sherry resigned herself to using a mechanical can opener. Then one day Ryan plunked her old can opener back on the counter, reassembled. Sherry appreciated the gesture, but doubted that her son, at his age, had done much more than stuff the entrails back inside and clamp it shut. "I don't think that'll work anymore, Ryan," Sherry said, but just to humor her son, she plugged it in and tried opening a can of tomato soup. To her amazement, it worked.

Meanwhile, the questions came nonstop. *Where does electricity come from?* "It comes from a power plant." *What's a power plant?* "It's a place where turbines go round and round." *What's a turbine?* Ryan's parents took him on a tour of a power plant, but even that didn't shut him up, prompting the Pattersons to start having questions of their own. What was driving Ryan to follow every electric current to its source? Once he hit a wall—which was often, metaphorically and literally—what measures would he take to get past it? Fearing that standing in his way might prove more dangerous than careful guidance, the Pattersons allowed their son's obsession to grow. One day, when an inspector came by to check that the wiring in a new room of the Pattersons' home was up to code, Ryan trailed him from socket to socket.

"Do they work?" Ryan asked. "Did I do it right?" Earlier that day, Sherry had sat her son down and said, "Now, Ryan, don't tell the inspector that Daddy let you wire the new room in the house." Thankfully, the inspector didn't take Ryan seriously. After all, he was only five years old.

At age eight, Ryan donned a baseball cap and shuffled into the outfield. From the bleachers, his parents cheered him on, praying Ryan would develop a taste for something other than outlets and extension cords. Yet again, no such luck. As a fly ball soared straight toward their son, he was staring at the ground, sketching shapes in the dirt with his feet, intensely absorbed in . . . something. The coach started screaming. Ryan's teammates started screaming. Even Ryan's parents started screaming. "Ryan?! RYAN?!" But it was no use. After the game, Ryan lingered in the outfield, seemingly unaware that the game was over, until Sherry walked out there and asked, "Ryan, what are you doing?"

"I'm drawing robots" was his response. Or in Ryan's case, *wobots*. Ever since he could speak, he had suffered from a slight speech impediment. It had been cute at age two, but at age eight, it was unfortunate. Since he was a shy kid, without many friends, robots made a sad sort of sense. Lacking friends, Ryan was designing his own.

One day, Ryan's dad, Randy, hopped into his pickup truck to find that the clock was missing, pried from the dashboard like a bad tooth. Next, the windshield wiper pumps disappeared. After that, Randy started finding notes taped to his steering wheel.

Please bring home one piece of steel
½ inch wide 16 gauge 12 inches long
Thanks dad
Ryan

Randy worked at a welding and fabrication company, so he was able to humor his son's odd requests. Randy also taught Ryan how to use an array of tools, from a soldering iron to a blowtorch, so Ryan could put his robots together. Working madly, like a child Frankenstein, at his miniature tool bench in the garage, Ryan fashioned heads, limbs, and torsos out of whatever he could get his hands on, from milk cartons to the insides of stereos. Finally, after months of work, he unveiled his first creation.

His name was Scorch. Scorch was dressed in a blue-and-orange-striped shirt that Ryan's grandmother had sewn for her grandson. The shirt was too big for Ryan, but it fit Scorch perfectly. Scorch had two red indicator lights for eyes, a row of five green incandescent lights for a smile, a remote control claw, and—thanks to the windshield wiper pumps Ryan had pinched from his dad's truck—a water cannon that could drench victims with pinpoint precision up to twenty feet away.

At Ryan's bidding, Scorch wandered the house, terrorizing his older sister Kim and all her friends. To keep family members from barging into his bedroom unannounced, Ryan installed an alarm on his door, as well as video camera surveillance in the hall. Ryan didn't like to be disturbed. This was especially true when he was reading the manual to an electronic device, from a Walkman to a cell phone, which he read religiously, cover to cover. His parents couldn't fathom how their son could relish reading something so tedious, but to Ryan, the manuals were as gripping as a mystery novel. Sadly, with people, the same desire to connect seemed oddly absent from Ryan. If friendliness were a switch, his was turned off.

By second grade, Ryan had become so reclusive that he begged his mom to let him stay home from school. Permanently. Why? Sherry asked. Was someone bullying him? Not yet. Were kids making fun of his speech impediment? Not yet. The reason he wanted to stay home, Ryan said, was to build robots. They needed him. He needed them. At school, one of Ryan's teachers sat Sherry down and broke the news. Ryan was bright. So bright, he might inevitably give up on school, give up on friends, give up on plugging into the world around him. The Pattersons had always known their son was different. But now it was time to panic.

• • •

At about this time, John McConnell was puttering around in his woodworking shop. At sixty-one, after decades of working as a physicist, John had retired from his post at Los Alamos National Laboratory. Now that his kids and grandkids were grown, he and his wife, Audrey, had moved to Grand Junction to breathe in the dry air, contemplate mountain vistas, and enjoy their retirement. So far, everything was going as planned. Occasionally, John wondered: *Is this all there is?* But these moments of unease were typical for new retirees, he'd been told. For the most part, he was content with the slower pace of life. Then he got a phone call from Sherry Patterson.

"I understand that you're a physicist and know about electronics," Sherry said. "I also understand you've done some mentoring." Recently John had started putting in a little volunteer time tutoring a student at Mesa State College. Sherry was hoping that John might like to mentor her son as well.

"How old is your son, ma'am?" John asked.

"He's in third grade."

Third grade? Mentoring college students was one thing. Babysitting was another. Still, something in this woman's voice, a tinge of desperation maybe, convinced John to drive over to the Pattersons', where Sherry greeted him with a grateful handshake. Then John and Ryan squared off with each other. At first, neither was all that impressed. *Wow, this guy looks older than Benjamin Franklin* was Ryan's first thought. *How can he possibly know about electronics?* Meanwhile, Ryan, who was small for his age, painfully shy, and still dogged by a slight speech impediment, struck John as terribly young. There was no way this would work.

"So, Ryan," John said to get things rolling. "I hear that you've created a robot. Can I see it?"

Ryan led John to the garage, where Scorch was waiting. After perusing Scorch's lightbulb eyes and grimy sweater, John tossed Ryan a question or two that he deemed appropriate for a boy his age.

"So what can Scorch do?"

Ryan ran Scorch through his usual tricks while John and Ryan watched and talked, much like guys might bond swapping stats on a vintage car. Slowly, John's questions became more detailed. What is Scorch's power source? How do the switches work? Like a car mechanic showing off his handiwork, Ryan obligingly opened Scorch up to display the wiring inside, then asked John a question of his own. Could John help make Scorch better?

As their allotted half hour ticked by, their mutual inquisition picked up steam. Ryan had never encountered a man who could answer all of his questions, so Ryan bored into him like a drill bit. John had never seen anyone, much less a boy, with such laser-like focus. In spite of the fifty-two years between them, John and Ryan clicked. Maybe the age gap between them made sense. Back in Los Alamos, John had been riffling through a magazine when a Winnebago motor home ad caught his eye. Its caption read: *Life is two periods of play interrupted by forty years of work.* John had liked the ad so much, he'd taped it above his desk. Now John had finished his forty years. He was ready to jump back into the sandbox and play, and Ryan was just getting started.

An hour flew by. Then two. Then three. Finally John stumbled back into the house, where Sherry had been hovering, wondering what was taking so long.

"Is everything okay?" Sherry asked, worried that John must be exhausted. But John was elated, electrified.

"How about I meet with Ryan every Saturday?" he asked.

John set aside wood turning. Since meeting Ryan, he has barely found time to carve one wooden bowl. Instead, he turned his woodworking shop into a classroom, where he and Ryan tinkered with semiconductors and integrated circuits. John quickly realized that an hour wasn't long enough for Ryan, who looked heartbroken whenever he left, so their hour-long sessions expanded to eight hours or more. At lunchtime, John dragged Ryan into the kitchen, and Audrey prepared them lunch, then dinner, then occasionally a bed where the boy could sleep as Ryan peppered John with every question he'd been stockpiling in his mind that week. John taught Ryan not only how electricity worked, but how to see the world through a scientist's eyes. "Don't think," John told Ryan. "Imagine." So Ryan dreamed of machines. In fourth grade, when a Valentine's Day box-decorating contest was announced at school, Ryan decided it was time to start building.

On February 14, amid cardboard shoe boxes decorated in construction-paper hearts, crowds gathered around Ryan's box, which wasn't just a box, but a metal fortress. That moved. And talked. As kids lined up to drop a Valentine's Day card through the roof, the windows lit up. A windmill turned. Then a tinny version of Ryan's voice kicked in, saying, "Thank you very much! Happy Valentine's Day!"

That day, Ryan received a lot of valentines. Which was ironic, since Ryan had few admirers. Science, after all, wasn't something that typically won kids many popularity points at school. Once middle school hit, even though his speech impediment had subsided, Ryan's eccentricities turned him into an easy target for bullies, who started coming on strong. At one point, Sherry was horrified to learn that Ryan had been beaten so badly he'd been sent to the doctor to ascertain if he'd suffered hearing loss. Ryan remained tight-lipped about these incidents, but his family could read between the lines. During Ryan's freshman year of high school, his sister Kim took it upon herself to invite her brother to sit with her at lunch every day to keep him company. Kim, unlike Ryan, was well rounded. She loved horses, softball, playing the flute in band, even weight lifting. Plus she had plenty of friends, a phenomenon that puzzled her little brother.

"How does Kim have so many friends?" Ryan asked his mom one day. "How do you get friends?"

Ryan's question broke Sherry's heart. "Well, you're nice to each other," she ventured. "You talk." But Ryan could only talk in terms of circuit breakers and voltage meters. And the only person who enjoyed talking about this stuff was John.

As mentor and protégé continued meeting every Saturday, John's wife, Audrey, noticed that the twosome had begun gesturing in the same manner and could even finish each other's sentences, picking up where the other left off. They were an odd pair, into odd things. But their relationship worked. *John looks like he could be my grandpa,* Ryan wrote in an essay for school, *and whenever we go somewhere, people assume that I'm his grandson. I like that a lot. I consider John McConnell my best friend.*

Since meeting Ryan, John had changed. The white-haired retiree was no longer winding down, but winding up. John wanted more. During the week, he and Audrey packed the trunk of their Honda with boxes filled with an odd hodgepodge of items, from Cheerios to tuning forks. Traveling from school to school, John and Audrey stood before classroomfuls of elementary school students and demonstrated the magic of science. John made Cheerios dance and twirl in midair, thanks to the magic of static electricity. He made spare change appear then disappear, with the help of a few lasers and a convex mirror. He blew bubbles in the shapes of cubes and pyramids. John became known throughout Colorado as the David Copperfield of science, and kids clamored for more.

John lobbied a local school for space, then filled it with more elaborate exhibits, from tanks full of waterbugs to human-sized kaleidoscopes. Later on, it would become known as the John McConnell Math and Science Center of Western Colorado. Kids from across the state came by the busload. All too soon, John, who'd come to Grand Junction to turn wood and enjoy his retirement, was putting in eighty hours per week for free. But John didn't mind. He was on a mission. Through mentoring Ryan, he had come to realize that catching kids when they're young was essential for igniting their interest in science. Wait too long, and the window closed, the spark drowned out by the chorus of kids who'd branded science uncool. Luckily, John had caught Ryan in time. And now, after years of getting picked on and treated like a pariah, Ryan's social life was finally looking up.

It started near the end of his freshman year in high school. Ryan was sitting in computer class. Behind him sat two seniors who were first-string players on the football team. Under normal circumstances, Ryan would have had no business saying a word to these guys, who were gods at Central High. But even jocks need to keep their grades up, and computers were their Achilles' heel. Right in front of them, Ryan was breezing through every assignment.

"Hey, Ryan," one of the jocks said. "Can you help us out?"

Ryan could tell from their tone of voice that they weren't merely trying to intimidate him to hand over the answers. These jocks were asking him to teach them the ropes. They needed Ryan. Ryan needed friends. The opportunity was ripe to strike a deal, so they did. While this reaching across the aisle of high school clique-dom seemed miraculous in itself, the jocks took it one step further. "Let's go to lunch," they proposed one day. Ryan's circle of friends grew from there. From then on, the bullies who had dogged Ryan throughout middle school steered clear. Which was good, since by this point, Ryan had bigger battles to fight, starting with science fairs. For that, he would need to turn back to his old pals: robots.

First, there was Mazebot. Mazebot was small compared to Scorch, no bigger than a lab rat. But Mazebot could do something amazing that Scorch could not. Thanks to photodiodes studding the sides of the android, Mazebot could sense walls and openings and reason its way through a maze, all on its own. At Colorado's state science fair, Mazebot won first place, qualifying Ryan for the Intel International Science and Engineering Fair 1999. At Intel ISEF, Mazebot won a small award, but Ryan wanted more and started mulling over what he could create next.

That spring, in nearby Columbine, Colorado, two students walked into school with shotguns to carry out the Columbine High School massacre. The news rattled schools across America. Colorado kids were hit particularly hard. At Central High, students had to evacuate due to a bomb scare. At the elementary school where Ryan's mother worked, perceived threats caused the school to be locked down several times. Many were haunted by what could have caused those kids to snap. Many blamed bullying. But amid the storm of negative news articles, there was a bright spot, thanks to Ryan Patterson.

Its name was Sleuthbot. Armed with sonar, infrared sensors, stereo hearing, and a video camera, this shoe box–sized device could search buildings for bombs and other threats at as brisk a clip as a cop. During Ryan's sophomore year, Sleuthbot won numerous prizes at Intel ISEF 2000. But even then, Ryan wasn't satisfied. During his junior year, Ryan set his mind to creating a robot that was even better than Sleuthbot, Mazebot, and Scorch combined.

Days later, Ryan and John were sitting in Burger King. Nearby, Ryan saw a deaf girl about his age who'd ordered her meal and was now chatting with friends with the help of an adult interpreter. Soon after that, Ryan stumbled across an article in the local newspaper about another deaf girl who confessed that it was incredibly awkward to have an adult interpreter hovering by her side all day, privy to her every comment. What if she were confessing to a crush on a boy? What if her girlfriends had a party planned when her parents were away? Ryan, who had once struggled to bond with his peers, could relate.

Ryan decided he would help this girl in Burger King, by creating a glove, a glove that could do something that no glove had ever done before. After hearing Ryan's plan, John drove down to the local sporting goods store and made his way to the golf aisle. He knew, based on countless rounds of golf he'd played through the years, that golf gloves were tough but flexible. They'd be perfect for the project at hand. Next, John and Ryan enlisted Audrey to sew ten flexible sensors into the fingers and palm of one glove.

These sensors could record movement, much like a Wii remote, and transmit that data to a tiny computer Ryan had attached at the wrist. That way, the deaf girl at Burger King could slip on the glove and sign letters of the alphabet in the air. The shapes her fingers made would be translated into text, which would pop up on a liquid crystal display she could hand to others. She could say, "A Whopper with cheese, please" directly to the Burger King cashier. She could tell her girlfriends, "Isn't that boy two seats down cute?" without an adult interpreter eavesdropping. The girl would be free from the hassles and humiliation that came with being different, all thanks to a glove that cost only about two hundred dollars to make.

At Intel ISEF 2001, Ryan's glove won $220,000. By the time he graduated high school in 2002, his winnings had topped $420,000. Media outlets from *People* magazine to CNN descended, clamoring for Ryan to sign something with his glove for the cameras. On CNN, he signed "Hi, Mom," which made Sherry melt. On *Good Morning America* in New York, Ryan was instructed to sign, "Good morning, Charlie" to the show's host, Charlie Gibson. But as Ryan was waiting to go on air live, he couldn't stop thinking about a phone conversation he'd had with John just minutes earlier. It wasn't just any old phone conversation. It was a man-to-man, heart-to-heart about a girl.

At school, after years of being seen as a nerd, Ryan had finally emerged as a wildly popular kid. Suddenly, as if they'd sprung up out of nowhere, Ryan discovered girls, and he discovered that he liked them, a lot. For the first time in his life, his productivity on the science front ebbed, as Ryan immersed himself in women. Currently, he was dating a girl named Tiffani, who was gorgeous and president of their class. For the past few weeks, he'd been hemming and hawing about asking her to the prom. Back at

Central High, Ryan knew that every TV in school would be tuned in to *Good Morning America,* and that Tiffani would be watching, and perhaps wondering what was taking him so long to pop the prom question. After hearing about Ryan's dilemma, John gave him some sage advice just minutes before he went on air: "Ryan, my boy, you only live once. Why sign 'Good morning, Charlie' when you could sign something you'd actually like to say?"

Live on *Good Morning America,* Ryan handed Charlie Gibson the liquid crystal display. Charlie read out loud, in front of millions of Americans, what Ryan had signed. "Prom? Tiff?" Charlie Gibson was confused. But Tiffani, back at Central High, wasn't confused in the slightest. Flying back to Colorado, Ryan was greeted at the airport by Tiffani, holding a cake, frosted to mimic his glove's liquid crystal display, on top of which was her answer: *Yes. Seventeen* magazine hailed Ryan as one of the hottest boys of summer. *Teen People* dubbed him one of the top twenty teens who would change the world. *GQ* devoted nine pages to detailing his genius. He met President George W. Bush, twice. Colorado's governor even proclaimed May 24 Ryan Patterson Day.

To top it off, Ryan used his copious science fair winnings to buy himself a really sweet car. Ryan's mom threw a fit the instant she laid eyes on her son's red-orange Mustang with a spoiler in the back. "But it's red," Sherry moaned. "You're going to get stopped by every cop in town." (Over time, though, Sherry grew to love the color and to drive the car herself at every opportunity. And Ryan never got a speeding ticket.) Ryan could easily have celebrated his new purchase by taking one of his entourage of friends for a spin. Instead, Ryan made a beeline for John's.

"Wanna go for a ride?" Ryan asked after pulling into John's driveway, kicking up dust. As the twosome cruised the neighborhood, they kidded around with each other as usual. Ryan and

John even had a running joke about the car Ryan would buy for John one day.

"One day, when I make a lot of money, I'll buy you a Porsche," Ryan said.

"You'd better hurry up," John warned. "Or else I'll be so old you'll need to buy a driver to go with it."

So much had changed in the eight years since John had first met Ryan and seen a glimmer of something special in a small, shy boy with a speech impediment, who spent the majority of his time with a robot named Scorch. What would Ryan's life have been like if he hadn't meant John? What would John's life have been like if he hadn't met Ryan? Neither of them liked to think about that too much.

Since Ryan would soon graduate high school, colleges came courting. Ryan allowed himself to be wooed, compiling a wish list of whims that might tempt him to commit. First, he wanted to skip core engineering courses that covered material he'd already learned in fifth grade. He wanted a suite rather than a mere dorm room, and a friend from Grand Junction would be joining him. Ryan would also need his own lab space where he could continue developing The Glove, a privilege typically unheard of for lowly undergrads. The University of Colorado in Boulder kowtowed to all his requests, so that's the college Ryan deigned to grace with his presence.

Ryan continued to dabble in dating in college, too, but it wasn't until after graduating that he met a girl worth pursuing in earnest. Most girls, Ryan complained to his mom, were too high-maintenance. But Marissa was different. At a party where they'd met, she broke the ice rather than waiting for Ryan to make the first move. She lived with her parents in Arvada, Colorado. And while Marissa didn't realize it yet, growing up, her parents had clipped and saved articles about Ryan that had appeared in the *Rocky Mountain News*. In the wake of Columbine and all the negative coverage about kids, Coloradans needed good news, and Ryan was it. Once Marissa's parents realized whom their daughter was dating, they greeted Ryan at the door gushing, cameras clicking to commemorate the event, thrilled that their daughter had landed such a catch.

Marissa was full of surprises for the Pattersons, too. During a trip by Ryan and Marissa to visit his parents, Sherry asked Marissa if she'd like to go into town and go shopping, figuring that's what girls liked to do. Marissa politely refused. Once Sherry got home that evening, she was surprised to find Marissa covered in dust, wielding a nail gun, helping Ryan and his dad with their latest renovation project. Ryan also introduced Marissa to John and Audrey, who said that they liked Marissa a lot. That was all Ryan needed to know. He was ready.

Ryan and Marissa drove to Copper Mountain to go snow-boarding, then spent the evening in front of a fire, sipping hot chocolate. Since it was cold out, Ryan was wearing gloves, and inside one glove, he had a surprise. They'd been dating for less than four months. He knew it was fast. But Ryan had learned to trust his hunches, thanks to John. Fifty-odd years earlier, John had met Audrey at a drive-in diner. After they'd dated four months, John proposed. Life was short. So when something felt right, you reached out and grabbed it.

Inside the glove, Ryan closed his hand around the ring, knelt on the ground, and pulled the glove off with a flourish. No, it wasn't quite as flashy as asking a girl to the prom on *Good Morning America*. Years later, Marissa would tease Ryan by asking,

"Why haven't you done something like that for me?" Someday, Ryan promised her, he would. It all depended on what he invented next.

During my tour of Grand Junction, it became clear that Ryan's legend lives on. At Central High, a trophy case displaying his science fair winnings stood side by side with the football and basketball players' cases for years. At his childhood home, Ryan's first robot, Scorch, remains holed up in the garage, wearing the same blue-and-orange-striped shirt sewn by his grandmother. The power outlet Ryan poked with a butter knife at age one still sports the telltale scorch marks.

"A few times we were going to replace it," Sherry offered as an explanation, "but then it reminded us of his childhood, so we don't."

Today, Ryan lives in Denver, Colorado, and works as an electrical engineer at Lockheed Martin, building spy satellites and spacecraft. He will never forget the difference science fairs made in his life. "I was a really shy kid," Ryan said. "What science fairs did for me more than anything was build confidence. I think science should be pressed a lot more in schools."

In addition to visiting the Pattersons, I swung by the John McConnell Math and Science Center, where John was wowing kids with crackling Tesla coils, square bubbles, and more than 150 other hands-on exhibits. Over the tanks of waterbugs, a neon sign glowed with a message that John hoped would stick with at least a few of the one hundred thousand kids who'd come through his doors: *Math and science are cool.*

Some kids, John said, were regular visitors, like Austin, a sandy-haired five-year-old whose mother typically brought along a peanut butter sandwich in case Austin's energy levels flagged halfway through. Hearing about little Austin, I couldn't help wondering if he was Grand Junction's next up-and-coming science fair star. Maybe one day, Austin's accomplishments on the science fair circuit will be commemorated with red-carpet rallies and a trophy case in school. Ryan and John blazed the path. Now all kids need to do is follow.

ELIZA AND THE BEES

Aerodynamically, the bumblebee shouldn't be
able to fly, but the bumblebee doesn't know it
so it goes on flying anyway.

—MARY KAY ASH, FOUNDER OF MARY KAY COSMETICS

Of all the competitors to grace the halls of the Intel International Science and Engineering Fair 2009, few seemed more puzzlingly out of place than Eliza McNitt. Consider her outfits, for starters. While most students dressed in business suits to convey how gravely they considered this undertaking, seventeen-year-old Eliza pranced around in vintage dresses, Valentino stilettos, and other designer duds courtesy of New York City's finest boutiques. During Intel ISEF, Eliza even managed to squeeze in some shoe shopping and a bikini-clad dip in the hotel Jacuzzi. In fact, if you transplanted Elle Woods from the movie *Legally Blonde* to a science fair, you'd have Eliza, right down to her long, platinum blonde hair.

The more I heard of Eliza's antics, the more convinced I became that I'd stumbled upon a rare breed of science fair star known as the anti-nerd. At age two, she'd started working as a Ford model. At age six, she'd kicked off her acting career as Dorothy in *The Wizard of Oz*. Eliza had also appeared on TV, playing a sick child in the soap opera *One Life to Live.* At sixteen, she'd landed a role as a girl crying in front of a police car in the movie *Life Before Her Eyes,* starring Uma Thurman. In addition to acting and modeling, Eliza was also a very good dancer. So good, in fact, that at one point during an evening mixer at Intel ISEF, her slinky, sinuous moves on the dance floor inspired a boy from Texas to state in awe, "I had no idea science fair girls could dance like that."

All of which begged the question: What was a girl like Eliza doing at a nerd convention like this? When I asked her this question, she laughed. "I got here by accident," she explained, before telling me the whole story.

Eliza lived in a mansion perched on a rocky outcrop, surrounded by the gentle swells of the Long Island Sound. Her house was painted pink and even had a name—Casa Chica—etched in a bronze plaque out front. After hearing that a Venezuelan ambassador with two daughters had once lived there, young Eliza took this to mean her home was once occupied by royalty and named her pets accordingly. Her cats were dubbed Duke of Dale and Duchess of Carbon; her goldfish was crowned King Lavish. To round out her animal kingdom she'd requested a flock of doves. Her parents refused, albeit reluctantly.

Eliza's parents had first crossed paths at a party. Audrey Appleby, a dancer, was twirling to the tunes of Marvin Gaye; Jim McNitt, a photographer, felt inspired to snap some pictures of her graceful, groovy moves. Audrey hoped to make it big on Broadway, but when that didn't happen, she shifted her aspirations toward opening a dance school for kids. Once Eliza arrived, Audrey found yet a new calling: She'd make her daughter a star.

When Eliza was two, her mom wheeled her stroller into the Ford Modeling Agency to drop off some photos of Eliza. She had barely gotten through the front door before a Ford executive spotted Eliza, asked, "Does this little girl have an appointment?" and whisked them upstairs. Within days, Eliza had booked her first photo shoot.

Eliza loved certain aspects of modeling—the bright lights, the amazing outfits—but as she got older, she resented the fact that she had no say in how she dressed, looked, or even moved in front of the camera. Like a doll, she had to passively accept other people's directions to smile, pout, and tilt her straw hat this way or that. *But what about what I want?* she wondered. Eventually, her attitude got her in trouble. One day when she was four, a photographer who was shooting her for a magazine snapped. "This isn't about you, Eliza, this is about getting the shot that we need. Stand still!" Soon after that, Eliza informed her mother that she was tired of modeling and wanted to be an actress. Audrey was thrilled and signed her up for a children's theater company.

At age four, Eliza landed her first acting role as a sheep in a production of *Lambert the Sheepish Lion*. On stage, she made a concerted effort not to glance out into the audience at her mother, who made pained facial expressions and silently mouthed Eliza's lines throughout her performance. Given that Eliza was too young to read, Audrey would rehearse her daughter's lines with her over and over until Eliza memorized them. Occasionally, when certain phrases refused to sink in, a mini meltdown ensued. But in the end, when she stepped on stage, Eliza was in her element. Her favorite actress was a toss-up between Audrey Hepburn and Ingrid Bergman. Her favorite movie was *Casablanca*, which she fondly remembered seeing with her family during their summer vacations in France.

In between her acting gigs, Eliza attended Greenwich High, a public high school bustling with nearly three thousand students. The school offered almost three hundred different courses, fifty varsity sports teams, and an even wider range of extracurricular activities, from the Anime Club to Dumbledore's Army (*Harry Potter* fans who'd formed their own nonprofit). At Greenwich High, it was easy to feel lost amid this banquet of options. To keep from slipping through the cracks, students typically found a niche and clung on for dear life. Well-rounded students rarely stood out; here, you picked a track and ran with it. There was the business track, the politics track. Eliza was squarely in the theatrical track. If she minored in anything, it was planning parties. As student government president, she arranged a variety of school-related social events, from hula hoop–offs to human foosball.

Eliza's home, Casa Chica, was also party central, and for every party Eliza racked her brain for a creative theme. One year, the theme was Eighties Workout. Eliza wore a pink unitard, leopard-print leggings, and scrunchies in her hair, while her classmates showed up in sweat suits, sweat bands, and really short shorts. For her sweet sixteen, the theme was *Alice in Wonderland,* complete with aperitifs labeled *Eat me* and *Drink me* and a Mad Hatter tea party. But the most original party of all was the Periodic Table Party, where attendants were required to dress up as one of the elements. On the big night, Eliza zipped herself into a neon blue body suit to represent tungsten, the glowing metal in lightbulbs. Answering the door, she was presented with her first guest: a classmate dressed as a pirate. After puzzling over it a bit, Eliza gave up. So what element was he?

"Arrr-gon," he said, before sauntering in.

Other classmates came dressed as a banana (potassium), a policeman (copper), and towing a bunch of balloons (helium). The occasion was that they'd all managed to pass chemistry, a class unanimously abhorred by all, especially Eliza. While she consistently made it into the top 10 percent of her class grade-wise, science never interested her all that much. It just seemed so dry, stuffy, unstylish, uncreative, and, well, *boring*.

But Eliza was about to be proven wrong.

During her junior year, Eliza was dismayed to find herself sitting in a science lab. Since her classes in acting and French occurred during the same periods as the more conventional science classes, the only course she could take to fulfill her requirement was hands-on research. Squinting through microscopes or pawing through frog guts was not Eliza's cup of tea. To make matters worse, everyone else in class knew it. Probably the most puzzled among them was a science nerd named Russel, who relished teasing Eliza at every turn.

"Hey, Eliza—what does HPLC stand for?"

Eliza, unfortunately, didn't know what HPLC stood for. There were actually a lot of things about science that Eliza didn't know. Nonetheless, she scrambled to come up with a decent retort.

"Nice shirt, Russel," she purred sweetly. Russel dressed the way most Greenwich guys dressed, in a wardrobe consisting strictly of Vineyard Vines Clothing. Preppy blue-striped button-downs, pink popped-collar shirts—in Eliza's mind, it was as if the Easter bunny had thrown up and turned it into a fashion line.

Their teacher, Mr. Bramante, or "Mr. B" as his students called him, was from the Bronx and had been teaching at Greenwich High for four years. While he'd managed to blend in somewhat by wearing the requisite polo shirt and chinos, occasionally he'd sport a rainbow tie-dyed lab coat. Mr. B also prided himself on having an impressive collection of funky glasses frames, which he tried to coordinate with his outfits. At five-eight, Mr. B was shorter than many of his students. He also had a receding hairline. The hair he had left was gray, even though every so often he got it colored, which prompted wisecracks from his students.

"My god, Mr. B, what did you do to your head?"

"I went to the hair salon," Mr. B explained. "They used the same stuff on me that they use on George Clooney." This would inevitably prompt another wave of wisecracks, not that he seemed to mind. Mr. B, unlike many of the teachers at Greenwich High, was laid back. Getting razzed by his students didn't ruffle him. Instead, he tried to understand where they were coming from. "What's with this texting business?" he once asked, then patiently listened as students explained. In lab, while students worked, Mr. B often plugged in his iPod and blasted classic rock, although he pointed out that while it might seem "classic" to kids their age, it was more accurately called eighties alternative. Occasionally he'd throw in a song or two that surprised them. Once, while he was playing the Benny Benassi song "Satisfaction," students rose up in protest. "That's *our* song," they exclaimed. "It's from the movie *White Chicks*." Mr. B had never seen the movie, but went out and rented it, and liked it. His open-mindedness made him that rare breed of teacher whom kids genuinely thought of as a friend. They invited him to their parties. Yet while Mr. B's demeanor was as mellow as it got, he also had the reputation for running one of the most grueling courses at Greenwich High.

In this class, students couldn't just kick back and memorize facts. Instead, each student spent the entire year doing science research. The time commitment required was so steep, in fact, that at the end of the school year Mr. B typically met with students who wanted to take his class the following year. That way, he could weed out kids who weren't all that serious about research and merely wanted to take his class because it fit conveniently into their schedule. These kids, Mr. B knew, would never survive the late nights in the lab, the painstakingly precise procedures, or the intense pressure that built, month by month, as science fair season approached.

But in spite of his efforts to screen his students, there sat Eliza, looking as out of place as a supermodel in the middle of a *Star Trek* convention. At Greenwich High, Eliza stood out pretty much wherever she went, and her wild outfits didn't help. One time, Mr. B was so puzzled by one of Eliza's ensembles, he had to ask what was up.

"Eliza, what's on your head?"

Eliza was wearing a hat in the shape of a bear. Yes, it was odd. And yet, in Mr. B's mind, odd could be good. He often encouraged his kids to think outside of the box when coming up with ideas for science fair projects, but rarely did students deliver. If anyone could throw him a curveball, it would be Eliza.

"So, Eliza," Mr. B said, sitting down with her to discuss possible projects. Other students, who'd met with him before summer, were already at work. Eliza was behind and would need to make up for lost time. "What would you like to do?"

Ever since she realized she'd be stuck in Mr. B's class, Eliza had been trying to wrap her brain around how to think like a scientist, whatever that meant. In acting class, she had learned how to inhabit any role, body and soul, and had successfully pulled off parts that were highly out of character. Once, she played a woman in an insane asylum. She'd even played a male role, as Shylock in Shakespeare's *The Merchant of Venice*. Getting into the mind-set of a scientist shouldn't be that much of a stretch. All she had to do was channel her grandfather.

While Eliza's parents were both artists, Eliza's grandfather was a chemical engineer, and a highly esteemed one at that. Back in the 1950s, he was asked to work for the U.S. government and was given a choice of one of two projects. One project focused on finding a way to mass-produce a new drug called penicillin. The other was top secret and had something to do with building a bomb, an enterprise that later became known as the Manhattan Project. Weighing that it would be better to help millions of people than hurt them, Eliza's grandfather picked penicillin. The gravity of his decision had always impressed Eliza, whose life seemed frivolous by comparison. So when he spoke, Eliza listened. She remembered how, one time, he'd told her to wash the apple she was about to eat, since it was probably covered in pesticides. As Eliza held the apple under the faucet, she wondered whether other things she liked to eat contained pesticides that might not be so easily removed. Scanning the kitchen counter, her eyes lit upon the jar of honey, which she used often in her tea. Could honey contain pesticides, too?

As Eliza babbled on to Mr. B about honey, tea, flowers, and the grandfather she admired so much, she could feel herself blushing. Clearly, these musings couldn't be worthy of a science fair project. But to her surprise, Mr. B seemed intrigued.

"You know, that's a pretty slick idea," he said. "That's really creative."

Hearing the word *creative,* Eliza perked up. Creative, she could do. Creativity was her strong point. Maybe science wasn't as outside her comfort zone as she'd thought. Still, a science fair project couldn't be built on creativity alone. Eliza needed evidence and a plan of action that could help her test her hypothesis. That would mean she'd have to venture far afield from her usual appointments, like acting auditions and shopping sprees in downtown

Manhattan. She would need to head into the wild and come face-to-face with the one place that bees called home. The hive.

Eliza breezed by peonies, petunias, magnolias, and rhododendrons at the Bartlett Arboretum in Stamford, Connecticut. Under normal circumstances, she could have whiled away hours admiring the beauty of her surroundings, but today she had more pressing priorities. At the center of this thirty-acre woodland stood the arboretum's beehive. Eliza could hear it long before she could see it, and the sound gave her the creeps. It sounded like a million violins playing at once.

Finally, in a clearing, Eliza spotted a whitewashed wooden box, on stilts, about the size of a small refrigerator. Slowly, she approached the beehive from the front. She quickly learned this was a bad idea when a bee shot out of the hive and nearly landed in her mouth. Before Eliza had time to recover, another bee came at her like a bullet. Then another. Without realizing it, she had parked herself right on the beeline, the path bees take to exit the hive. And clearly, she had caught them during their morning commute, since bees were shooting out of the hive about once every second, like well-timed flights at a busy airport. Given that these bees had a busy day of nectar collection ahead of them, they were no happier to encounter Eliza than she would have been encountering bumper-to-bumper traffic on her way to an important audition. As soon as Eliza realized her faux pas, she switched to approaching the hive from the side. At least, she was relieved to see, the bees seemed to have no interest in making her pay for her mistake. Even when the beekeeper who had escorted her to the area pulled out racks of waxy honeycomb, the bees couldn't be bothered to rise up in protest. They had more important things to do.

Eliza, like most kids, had grown up terrified of bees. When she was four, on a boating trip, a friend of hers got stung. The girl screamed, then sobbed as a flurry of family members came running. Boating was canceled. Eliza was shaken. And yet, perhaps because she expected the worst, when a bee stung her the following summer at camp, she didn't think it was all that devastating. Bees, she concluded, weren't as bad as their reputation suggested. They were just misunderstood. Given how people saw her—blonde, beautiful, actress, model, but not necessarily smart—she could sympathize.

The more Eliza learned about bees, the more fascinated she became. She swooned when she learned that a beekeeper operated a hive on top of the Paris Opera House, and vowed she'd one day pay him a visit. She was also impressed to learn that bees didn't just pollinate flowers and produce honey, but were responsible for pollinating crops that led to one out of every three bites of food we ate. Without bees, much of the world's plant life would wither. Apples would cost $15 apiece. The world would become a barren, ugly, unlivable place. Eliza had underestimated the bees, just as the kids in her science class had underestimated Eliza. Or so she hoped.

Eliza left the Bartlett Arboretum that day with a bottle of honey as well as small samples of the five pesticides the arboretum's gardeners sprayed on the area. She had conquered her first challenge. She'd even braced herself to get stung, but had escaped unscathed. Still, Eliza knew that this was just the first step.

Back in Mr. B's lab, Eliza approached an instrument that would become the bane of her existence: the HPLC. Eliza had learned that HPLC stood for high pressure liquid chromatography, but that didn't help her feel any less lost when standing before it. The HPLC looked a lot like a beehive and was about the same size. It also hummed like a hive. The instrument's exterior was covered in buttons, lights, and readout screens. Open its door, and inside was an intestinal mess of pumps, tubes, and wires. An HPLC, Mr. B explained to Eliza, could detect the chemical composition of a substance based on the arrangement of its electrons, or its polarity. Eliza didn't know exactly what that meant. All she knew was that if she was going to figure out if any pesticides were hiding in her honey, she would have to learn how to master this machine.

Thanks to Mr. B, Greenwich High was fortunate to have a lab that looked like it came straight out of a *CSI* episode. That's because before becoming a teacher, Mr. B had worked for Perkin-Elmer, a company that engineered equipment for forensics labs. Even though Mr. B had since turned to teaching, he kept in touch with his old work buddies, so whenever one of their clients decided to upgrade, Mr. B could scoot in and take their old instruments off their hands. As a result, Mr. B's lab was filled with instruments that could pinpoint whether poison was lurking in someone's food, or which chemicals laced a strand of hair. His lab was better equipped than many professional laboratories, and a lot of machinery in his charge was worth tens of thousands of dollars. All the more reason for Eliza to feel intimidated.

So far, Eliza's foray into operating lab equipment had elicited plenty of snickers from students. Once, she spilled insect-killing soap all over Mr. B's Fourier transform spectrophotometer, an instrument that cost more than a car. Gobs of honey also ended up everywhere—on tables, chairs, even in Eliza's hair. Even though she once loved having honey in her tea, she started drinking her tea without it. Due to her blunders, Eliza was nervous about getting anywhere near the HPLC. But by this time, it was January. If Eliza didn't get cracking, she wouldn't finish in time for the science fair. It was now or never.

As Eliza placed her honey sample inside the machine, she prayed that the chromatogram, or visual readout of the results, would show the telltale peaks and valleys indicating that for once, she'd done something right. And yet, within minutes, the HPLC machine started spitting out gibberish, which meant that another day's work was down the drain. Mr. B did his best to stave off student meltdowns by remaining encouraging. "Don't sweat it, Eliza," he said. "We'll just try it again."

Eliza tried. Day by day, week by week, she got the hang of the HPLC. She also got the hang of analyzing her data and writing up her results. At this point, all she needed to finish her research was a flower known as the pyrethrum daisy.

At the center of a pyrethrum daisy lay a natural insecticide called pyrethrine. Eliza's research wouldn't be complete unless she analyzed her honey for this substance as well. She called florist after florist, nursery after nursery, but no one had heard of the pyrethrum daisy. Days passed, then weeks. Eliza started to panic. She pored through the literature, looking for information about where she could find this rare flower. Then one day as she was surfing the Internet, it hit her: Pyrethrum was just a fancy word for your garden-variety daisy. Eliza felt ridiculous. She'd gotten so wrapped up in thinking of science as some complicated endeavor that she had overlooked an important lesson: Simplicity was the key to science. Eliza walked across the street to her neighborhood florist, bought a bouquet of daisies, buried her nose in them, and breathed deep. It had been a while since Eliza had bothered to do that.

• • •

By the time science fair season rolled around, Eliza was ready. Her research had found that honey did indeed contain pesticides, which could be toxic and a cause for concern among environmentalists. She looked forward to presenting her findings to the judges, figuring that a presentation was a lot like a performance. But the judges, she learned, were a tough crowd to please. Instead of kicking back and enjoying the show, they heckled her with questions about her research. "How do you know that high pressure liquid chromatography was the best method to use?" "How do you know that your readings are accurate?" One judge even asked her, "Who helped you conduct your research?" Translation: There's no way Eliza could have done this herself. After all, HPLC was a difficult procedure involving sophisticated equipment. Eliza was just a high schooler, and a girl, and a good-looking blonde to boot.

Sensing their skepticism, Eliza could have easily started backsliding. Instead, she thought back to the hardest role she'd ever tackled, in a play by Tennessee Williams called *Suddenly, Last Summer.* Eliza played Catherine Holly, a girl who witnessed the bizarre death of her cousin through cannibalism. Even though Catherine was telling the truth, no one believed her. They assumed she was crazy, and threatened to give her a lobotomy. Even then, Catherine stood by her story. To prepare for the role, Eliza had spent months trying to understand why someone who was threatened with a lobotomy would have the strength to stick to her guns. Eventually, the epiphany came to her: Catherine didn't waver because she had truth on her side. And Eliza did, too.

Eliza held her ground against judge after judge. At the awards ceremony, she took her seat alongside Mr. B and the rest of her classmates. When her name was called for second place, all the other students who'd been ribbing her all year applauded madly. Even Russel—preppy, pastel-wearing, tennis team captain Russel—gave her his congratulations. While Eliza hadn't suspected it, she had won over her classmates long ago.

Weeks later, Eliza flew to Atlanta, Georgia, to compete against more than fifteen hundred students at the Intel International Science and Engineering Fair 2008. Mr. B volunteered to go as well, acting as chaperone to Eliza and the other finalists from Connecticut. Chaperones had the rather inglorious task of chauffeuring the students to and from events and mixers. Since adults weren't allowed in, Mr. B typically waited in the lobby. Occasionally Eliza and the others would stay so late, he'd fall asleep, his wallet splayed out on the table in front of him (he hated carrying his wallet in his back pocket), until the students woke him up so he could drive them home. It was during that trip that Mr. B came to realize that Eliza was unique in the science fair world. While he couldn't say for sure what would happen, he had a feeling Eliza might have a shot at standing out.

Mr. B was right. Even at Intel ISEF, Eliza's star shone. At the awards ceremony, Eliza won second place in the Environmental Management category. When a judge or two asked her what she planned to study in college, their faces fell when Eliza said acting. "You should consider going into science," they said. "You're good at it. Science needs more people like you." Eliza was surprised. Up until now, she'd considered her accidental detour into the world of science fairs as just that: a detour, nothing more. Art—not science—was her true passion and path in life.

Next year, Eliza would be a senior. While she'd been thinking it would be fun to sign up for Mr. B's class a second time, she knew that wouldn't be wise, given that she'd have her hands full with college applications and auditions. In life, people had to make choices. And science and art just didn't mix. Soon, Eliza would have to say good-bye to the bees, to the HPLC and the Fourier transform spectrophotometer, to Russel's snide remarks, and last but not least, to Mr. B himself. Soon, her love affair with science would have to end.

"Whewww! What's that smell?"

Eliza was a senior, back in Mr. B's class after all, and she had just opened a bag of dead bees. Even though she was a seasoned science fair veteran at this point, she still had a knack for making a scene at her own expense, and today was a perfect example. Eliza, who'd acquired the dead bees from a local beekeeper who tended hives for Martha Stewart, had been too polite to ask how long the bees had been sitting in the bag. But based on the smell—similar to spoiled fish, only worse—it was clear they'd been sitting in there far longer than Martha Stewart would ever have deemed appropriate.

Within seconds, the entire lab was filled with a stench that sent Mr. B's latest crop of students fleeing en masse toward the door. Mr. B, who'd gotten the first whiff, since Eliza had triumphantly opened the bag under his nose, raised his eyebrows in surprise and pronounced that this was the worst thing he'd ever smelled (and for him, that was saying something). To keep this suffocating odor from permeating the lab, he introduced Eliza to a new device called the hood, an enclosed area that kept hazardous fumes from spreading. Eliza's bees weren't hazardous, but they might as well be, given how quickly they could clear a room.

These dead bees, Eliza had learned, were just the tip of the iceberg. Around the world, honeybees were disappearing in droves.

In certain areas, scientists estimated that 80 percent of the honeybees had vanished within six months. In Croatia, 5 million bees disappeared in less than forty-eight hours. In 2006, this phenomenon obtained a name: Colony Collapse Disorder (CCD). More baffling still, no one knew what was causing it, although scientists were piecing together a few possibilities. Some blamed a virus called the Israeli acute paralysis virus. Others claimed that urban areas had infringed on honeybee territory. While reading up on CCD, Eliza found evidence that pesticides might also play a role, given that the first case of CCD coincided with the introduction of a pesticide called imidacloprid to the environment. Curious to get to the bottom of this mystery, Eliza had managed to squeeze Mr. B's class into her schedule. She was going to test bee corpses for traces of imidacloprid, to see if this toxin might be the culprit.

If Eliza thought last year's research was rough, this year was downright disgusting. Now, instead of dealing with honey and sticky hair, she was dismembering bees part by part, plucking wings and legs from thoraxes, then the six little bee feet from the legs. From there, each bee part was thrown into a food processor. The sound of bee bodies being crushed, followed by the squirt of bee innards bursting out, was as gross as it got. Hood or not, few students could stomach working near her.

When not whipping up samples of bee sludge, Eliza was also taking a crack at something new: getting behind the camera and directing a documentary film. The idea had come to her in English class, thanks to a guy named Charlie Greene. Charlie and Eliza shared a special bond. They were born on the very same day, a mere two hours apart. Both of their moms had scheduled C-sections—Charlie's first, Eliza's second—and had passed each other in the delivery room like ships at sea. Since then, both families had bumped into each other at the endless parade of picnics and gatherings that characterized life in Greenwich, while Eliza and Charlie ran off to weave their own worlds of make-believe. Due to the odd coincidence characterizing their near-simultaneous births, Eliza and Charlie grew up regarding each other as twins. Now, as high school seniors, their lives were about to converge again.

Charlie, like Eliza, was interested in film, although their tastes couldn't have been more different. Eliza thrived on old French cinema. Charlie preferred more contemporary fare like *Donnie Darko.* Eliza loved to star in films. Charlie preferred to create them. He couldn't remember a time in his life when he wasn't holding a video camera, making his own home movies, or puttering around with editing equipment. After hearing that C-SPAN was accepting entries for a documentary film competition, Charlie set his sights on winning that. The theme of the competition was for contestants to call attention to an issue that they felt that President Obama needed to address. Charlie had considered all the usual options—welfare, the economy—but nothing seemed quite right. He needed something original. Just as he and Eliza were talking at the end of English class, the bell rang. For Eliza, the light-bulb went off.

"Colony Collapse Disorder!" she shouted to Charlie, even as the crowds started pulling them in different directions. *What's that?* Charlie wondered, then Wikipedia'd it later that night. He agreed with Eliza. It was definitely original. Days later, they were in the car together, camera equipment in back, heading toward Lewisburg, Pennsylvania. There, they'd film not just one hive, but three thousand hives that comprised Hackenberg Apiaries, one of the largest bee farms in the country. For the first time, Eliza found herself behind a camera instead of in front of it. And to her surprise, she loved it.

After shooting hours of footage, Charlie and Eliza went back to his place to begin the editing process. As the deadline for the C-SPAN contest approached, they sat side by side for hours, picking frames and dubbing in dialogue. Over Christmas break, they spent three days straight editing, while Charlie's mom brought them eleven consecutive meals. Eliza started borrowing Charlie's clothes. Occasionally, she got on Charlie's nerves by lapsing into speaking French or purring like a cat, which creeped him out. As their constant proximity reached breaking point, even the way Eliza breathed occasionally prompted Charlie to say, "Okay—we're done for today."

And yet there were also plenty of moments that, looking back later, would make Charlie and Eliza smile. Once, after she had gone home, Charlie stayed up late reviewing B-roll she had taken of flowers fluttering in the wind, clouds moving across the sky, and finally, a sunset. As Charlie stared bleary-eyed at the footage, he could swear he heard Eliza humming a Beatles song. Correction: *singing* a Beatles song at the top of her lungs. Even though Eliza didn't know all the words, that didn't stop her; she just sang the stanza she did know over and over. Later on when he confronted her, Eliza swore she had no recollection of humming or singing anything. After all, B-roll was just for visuals and would be paired with its own soundtrack later. Still, even behind the camera apparently, Eliza was a ham. Self-expression was her raison d'être. Later on, Charlie would text Eliza, *I'm watching the sunset, wish I had a soundtrack*. It became their personal inside joke.

Later on, they decided to go to their senior prom together. "We're just going as friends," they insisted to nosy family members. Even though Eliza could have easily had her pick of many guys at Greenwich High, she'd shied away from dating; the guys were too stuffy for her tastes. Charlie, however, was different. So when she decided to go to prom in something a bit unusual— a sea green sari-style dress from the 1960s—she knew Charlie would understand. When he saw her, he was more than understanding. He loved it.

Eliza and Charlie's documentary, *Requiem for the Honeybee*, won first place in the C-SPAN competition. At Connecticut's state science fair, Eliza's research on imidacloprid's role in Colony Collapse Disorder also wowed the judges, who selected her as a finalist for the Intel International Science and Engineering Fair 2009 in Reno, Nevada. Still, even at this point, Eliza had no plans to continue her work in science once Intel ISEF was over.

"When I'm at a science fair, I sort of feel like an outsider," Eliza confessed. And besides, she had her acting career to think of. As much as she loved science, Eliza had chalked it up to a fling rather than a long-term relationship. Soon, she would have to leave science behind.

Then something happened at Intel ISEF that changed her mind.

LISTENING TO LORENA

> *Most people never listen.*
> —ERNEST HEMINGWAY

Like most children with autism, Lorena Picone had her quirks. She shied away from eye contact. When she spoke, she referred to herself in the third person: *Lorena wants to go out and play.* Loud noises, like the blender, made her cover her ears and scream. The sound of certain words, like *Snoop Dog,* made her chuckle, so she repeated them over and over, a phenomenon known as echolalia. By age six, the only letter of the alphabet she had learned was the letter *L.*

At family gatherings, even when Lorena was surrounded by people she knew, she was essentially alone, her face expressionless, ensconced in her own private party of one. In the same way that Lorena was largely oblivious to the subtleties of her surroundings, most of her relatives had learned to tune out Lorena's idiosyncrasies, much as they'd ignore the drone of a TV. But Lorena's older cousin, Kayla Cornale, was fascinated with Lorena from the very first moment they met. *What is Lorena thinking?* Kayla wondered, as her little cousin wandered away from her mid-conversation, toward the piano, to bang a few keys and chortle. *What is Lorena trying to say?*

Kayla started listening. Bit by bit, she started hearing the first few whispers of Lorena's answer. It wasn't just noise. This revelation led to a science fair project that would rack up more than fifty awards, including top honors at the Intel International Science and Engineering Fair in 2005 and 2006. After catching wind of her story, I tracked Kayla down at Stanford University, where she was a sophomore. The lesson I would learn from her was that the ultimate reward of doing science fairs isn't fame, or money, or college scholarships. It's far simpler than that. It's about connecting with the people you care about most.

Kayla's story began in 2003, on the second Monday of October. While this date doesn't mean much to Americans, it does to Canadians, since that's when they celebrate Thanksgiving. Aside from their different dates, Canadian and American Thanksgivings are largely the same, and the Cornale gatherings were as classic as it gets. Every year, Kayla and her family would drive to Uncle Michael and Aunt Carolyn's home in Dundas, Ontario, and bask in a blur of turkey, gravy-laden stuffing, pies à la mode, movie rentals, and board games in front of the fire. But Kayla's favorite part was catching up with her cousins. There were thirteen of them, all younger than her, and Kayla considered it her duty to watch over the brood. At any point, her voice, which was surprisingly raspy for her age and reedy build, could be heard asking one of them what they'd been up to since the last family get-together or what they were working on in school. While all of Kayla's cousins soaked up the attention, few flowered under her gentle questioning quite like Lorena.

Lorena's parents had first started noticing there was something different about Lorena at around age one, when her ability to talk, walk, and interact with others seemed stunted compared to that of other kids her age. At age five, she was diagnosed with autism and placed in special ed. At family gatherings, her odd behavior sometimes ruffled her cousins or caused minor spats among the adults. If Lorena didn't clear her plate after dinner, relatives swooped in to clear it for her, while Lorena's parents insisted that their daughter could and should do it herself—just because she was disabled didn't mean she should be coddled. It also pained the Picones to hear about how well many of their nieces and nephews were doing in school—all thirteen would make the honor roll—while Lorena was struggling to learn the alphabet and stuck on the letter *L*.

While Lorena's attention span for most activities was sorely limited, certain things could capture her attention for hours on end. For one, she was fascinated by Kayla's long, straight blonde hair and could spend hours braiding it, which Kayla gamely allowed her to do. Lorena was also riveted by music, and her tastes were eclectic, ranging from rap to country. She didn't dance to it or sing to it, she just listened intently and laughed occasionally, although no one could fathom what was so funny. Uncle Michael mentioned how once, he sang Lorena a song that, to his surprise, she sang back to him a year later, having memorized every note and lyric by heart. Hearing this, Kayla was intrigued. Given that Lorena had only managed to memorize one letter of the alphabet, it struck Kayla as remarkable that her little cousin could memorize an entire song that fast.

Soon after that Thanksgiving gathering, Kayla was discussing Lorena's affinity for music with her family over dinner when something in her mind clicked: Maybe Lorena learned *through* music. If that were true, then maybe all she needed to learn the alphabet was to have it presented to her in a more musical manner. Only how could that be done? Even if Lorena could be taught to sing her ABCs, that wouldn't mean she understood what she was saying. Lorena needed a more hands-on method for the letters to sink in. Kayla's mind turned to the piano in Uncle Michael's and Aunt Carolyn's living room. Like braiding Kayla's hair, pounding the piano keys never ceased to amuse Lorena. Maybe she could be taught to associate certain notes with certain letters of the alphabet. It was a stretch, but given that Lorena's efforts to learn the alphabet had been stuck in limbo at *L* for a while, there seemed to be little harm in giving it a try.

Kayla discussed the idea with Lorena's parents, Greg and Alisa, who were happy to have Kayla come by on Sundays. Since Kayla was fourteen at the time and too young to drive, her mother, Mary, drove her there, then sat in the kitchen drinking coffee and catching up with Alisa as they kept an eye on Alisa's two younger daughters. Meanwhile, Kayla led Lorena into the living room, where the Picones kept their piano. Kayla had taken piano lessons from fourth to ninth grade, so she knew her way around a keyboard. But today she'd be trying something different. Sitting down, Kayla taped the letters *A* through *Z* on the white enamel of twenty-six keys, with the most common letters like *E* and *T* near the middle and less common ones like *X* and *Z* along the outskirts. Then she invited Lorena to sit down next to her.

"This letter is *A*," Kayla said, pressing the piano key with the appropriate sticker. "Lorena, can you play *A* for me?"

Lorena cocked her head to the side—a typical gesture when she was trying to suss out something new. Usually she loved poking at the piano keys. But today she recoiled. No matter how gently Kayla coaxed Lorena to place her hands on the keys, she refused. Perhaps the stickers scared her. Or perhaps she disliked being steered toward certain notes, preferring instead to plunk whichever keys struck her fancy. For whatever reason, Lorena wanted no part of this experiment, and unless Kayla could figure out why, that would mean she had reached a dead end on her very first day.

The minutes ticked by as Kayla and Lorena sat there in a silent stalemate, with Lorena staring off into space. Even though Kayla was tempted to throw in the towel, she wasn't willing to call it quits just yet. Part of this had to do with what Kayla knew about Lorena, based on the many Thanksgivings, Christmases, Easters, and other holidays they had spent together. Somewhere in those countless hours of hanging out, braiding hair, and listening to music and each other, Kayla had seen flashes of startling intelligence in Lorena. Even at age six, with autism, Lorena had a reputation for playing tricks on people, pretending not to know an animal in a picture book then turning around and naming the animal with ease once it suited her interests. Based on snippets like this that Kayla had amassed over the years, she had a hunch that, autism or not, Lorena was smart. She just lived on her own wavelength, which threw a wrench into her ability to smoothly transmit her intelligence to others. Maybe what she needed, then, was for someone to try tuning in to her frequency. Lorena needed someone to learn Lorenglish, and Kayla was willing to give it a shot.

Kayla didn't give up easily. At five feet, she was the shortest member of her high school basketball team, a sport she adored in a family who held sports in the highest esteem. Kayla's dad had been a high school quarterback, and Kayla's brother Cody had followed in his footsteps. On TV, it didn't matter if it was the NFL or CFL, pro or college, March Madness or Super Bowl Sunday. If sports were on, the Cornales were watching and whooping through every three-pointer and touchdown pass. At school, Kayla played five varsity sports, but basketball was her favorite. During games, Kayla was used to hearing members on the opposing team shout, "I've got the little one!" since she appeared painfully easy to guard.

Since she fell short on the height front, Kayla went all out to make up for it. She practiced dribbling drills in the driveway until the ball moved so fast it was a blur under her feet. She even played keep-away with her dad and older brother Cody, who was six foot four and often joked, "Don't make me come down there and beat you up." Once, when Kayla clocked him surprisingly hard during their roughhousing, Cody responded calmly with "I could clobber you if I wanted to, but I'm not going to." Cody was as benevolent as older brothers came, and dealing with thirteen younger cousins had only served to augment his already vast reservoirs of patience. Kayla, as the second oldest cousin, also felt a sense of responsibility toward her younger family members. Lorena, in other words, wasn't the one who needed to change. Kayla was. The onus was on her to find a way to break through.

As Lorena sat cringing before the keyboard, Kayla considered her options. Fine, so Lorena wasn't willing to touch the keys, at least not yet. What could Kayla do to entice her to give it a try? Kayla knew that Lorena looked up to her, so perhaps if Kayla played the piano, Lorena would warm up to the idea. So Kayla placed her hand on the keyboard and played A-P-E, acting as if she were immensely enjoying herself, as if to say, *See? This is nothing to be afraid of.* Once Kayla's piano playing appeared to pique Lorena's interest, Kayla inched to the next step. "Why don't you put your hand on top of mine as I play?" she asked Lorena, who tentatively complied.

For three weeks, Kayla played the piano with Lorena's hand sitting on top of hers. By week four, Lorena had grown comfortable enough to start touching the keys and playing *A-P-E* on her own. Even then, though, Kayla couldn't say for sure whether her lessons were sinking in. Was Lorena aware that she was spelling the word *ape*, or was she merely copying Kayla's movements? Then one Sunday, as Kayla sat down at the piano and waited for Lorena to join her, Kayla ran her fingers over a random sequence of notes, which happened to be *A-P-T.*

"That's wrong," Lorena immediately piped up from across the room. "That was *T*. You need *E*."

Kayla looked up, surprised. On the piano keyboard, *T* and *E* were right next to each other. Only a sophisticated ear would have been able to tell them apart. But Lorena could tell the difference with no musical training whatsoever. What other talents did her little cousin have under wraps, untapped? More importantly, by correcting Kayla, Lorena had proved that she wasn't merely parroting Kayla's movements. Instead, something far more interesting was happening: Lorena was learning. Just three weeks earlier, Lorena had known only one letter of the alphabet, the letter *L*. Now it appeared that she might know four more: *A-P-E* and *T.*

Sunday after Sunday, Kayla challenged Lorena with new letters, new animals, and new collections of notes. *B-E-A-R. C-A-T. D-O-G.* To test whether Lorena's grasp of these letters and words extended beyond the confines of the piano, Kayla led her to the dining room table and set out animal stickers and pieces of paper on which the corresponding words were written. When Kayla asked Lorena to match up the stickers with the words, she did so with ease. The Picones were delighted when, while reading to their daughters at night, Lorena began pointing out letters she recognized. By spring, she had memorized the entire alphabet and was reading certain words on her own.

As Lorena was mastering the world of words, Kayla was getting in the swing of her high school science fair. She had competed in her first fair in eighth grade, building a robotic arm. It was fun, but she was a jock, not a science geek, and had little expectation of winning. As she sat in the audience and heard her name called for first place, she was so shocked she swallowed her gum. Kayla headed home with $50 in prize money, with which she splurged and bought herself a new pair of basketball sneakers.

Now in ninth grade, Kayla was ready to give science fairs another whirl, and she had the perfect project: Lorena. This year, she came home with a slew of awards, qualifying as a finalist for the national science fair in Canada. Whenever judges asked what inspired her project, Kayla didn't hesitate. Her cousin Lorena, she explained, was her muse. But Lorena was more than just that. Lorena was also her buddy. Kayla might have opened Lorena's eyes to the world of words, but Lorena had opened Kayla's eyes as well. Before this point, Kayla had always seen her little cousin as a kid with autism. But Lorena had given her a peek at the person behind the condition. And Kayla was determined to know more.

Kayla and Lorena continued to meet every Sunday for the next four years. In spite of basketball, badminton, softball, volleyball, golf, and spending time with friends her own age, never once did Kayla view these Sunday sessions as a chore. In her opinion, it was a privilege to get to know Lorena. Autism made their relationship more challenging than most, but her little cousin was well worth the effort. As for those who weren't willing to make the effort, that was their loss.

As the subtleties of Lorena's personality slowly emerged, Kayla was surprised to find how much they had in common. For one, they had the same taste in music. Shania Twain, by far, was their favorite. Kayla also discovered that they even had the same taste in clothes. One Sunday, when she showed up at the Picones' front door wearing a green-yellow-and-blue-striped shirt she'd recently bought on sale at the Gap, Lorena opened the door wearing the exact same shirt. "Lorena! Look at us!" Kayla exclaimed, pointing to her shirt, then to her cousin's smaller version, which sent them both into hysterics.

As Kayla delved into her work with Lorena, she also dug into the available research. Kayla learned that the term *autism* was coined in 1911 from the Greek word *autos,* meaning "self," to describe the social isolation its sufferers often feel from the world around them. Kayla was surprised to find that today, 1 out of every 150 kids is diagnosed with autism, making it more common than childhood cancer, juvenile diabetes, and pediatric AIDS combined. Autism awareness was at an all-time high, thanks to celebrities with autistic kids, like Jenny McCarthy. But in spite of its prevalence, the exact cause of this neurological condition was unknown. Most said its roots were largely genetic. Others blamed childhood vaccines. Still others theorized that autism rates had always been high, but that more kids were being diagnosed with it, rather than lumped in with the learning disabled.

There is no known cure for autism. A variety of treatment options were available, but Kayla wasn't all that impressed. In one of the most common therapies, applied behavior analysis, kids with autism were presented with a spoon, fork, or cup and asked to name the object. If they did, they received a "reinforcer," like candy. To Kayla, this type of training felt rigid and repetitive. Kids with autism weren't robots. They were just trapped inside their minds, looking for a way out.

Kayla read that many autistic kids, much like Lorena, had an affinity for music, which experts attribute to their strong inclination for creating patterns and their hypersensitivity to sound (which would also explain why the blender made Lorena scream). Music, then, was like a language that many kids with autism understood, and it could serve as a bridge to learning other things, like the alphabet and words. That certainly seemed to be the case for Lorena, at least. But would it work for others? Curious, Kayla decided to find out.

Kayla rolled out her program, called Sounds into Syllables, in twelve special ed classes in nearby schools. She explained to teachers how they could implement her program with their own autistic students, asking in return that they monitor each student's progress for a pilot study. One day, Kayla got a call from a teacher at Holy Rosary Elementary School. "We have a first grader named Nathan who's really excelling," the teacher said enthusiastically, encouraging Kayla to come down and see for herself.

Just six months earlier, Nathan had only known the letter *N*. Now, thanks to Kayla's program, he'd mastered the entire alphabet, could spell every animal in her book, from *ape* to *zebra,* and was now responsible for writing the date on the class chalkboard every day. Nathan, while clearly a standout, wasn't the only child who was flourishing, either. Of the eighteen kids in her pilot study, Kayla learned that the majority were coming out of their shells, recognizing letters and spelling words for the very first time. Now the only question teachers and parents were clamoring for Kayla to answer was this: What could she teach them next?

Soon after that, Lorena came down for her Sunday session to find herself facing a far different instrument: a laptop computer. After taking a programming course at school, Kayla had written a software program that caused each computer key to play a different piano sound when pressed. Up on screen, instead of animals, Lorena was presented with a whole new set of words: verbs. For Kayla, moving her program to a computer made sense not only because computers were far more plentiful than pianos, but because they allowed her to incorporate animation, which made action words like verbs easier to portray. Since Lorena was leery of new things, Kayla tried to ease the transition by pairing each verb with an animal Lorena had already learned, like *frog* with *leap*. Once Lorena successfully spelled *leap* on the computer keyboard, the frog on the computer screen leapt in the air. It didn't take long before Lorena was laughing at leaping frogs, barking dogs, and roaring lions.

Next, Kayla set about teaching Lorena something far more complex than nouns, verbs, or words of any kind. Kayla wanted to see if she could teach Lorena emotions. Through her research, she had read that one reason kids with autism struggled with social interactions had to do with an area of the brain called the fusiform. The fusiform played a role in facial recognition, and typically lit up on brain scans of individuals as they gazed upon photographs of familiar faces. Only when autistic individuals were shown the same photographs, the fusiform remained dormant, like a burned-out light. Facial expressions, then, were a language that kids with autism couldn't easily recognize. But was there a way around this? Soon after that, Kayla stumbled across a study in which autistic individuals were shown cartoon faces. To Kayla's surprise, brain scans indicated that these simple line drawings did get their fusiforms glowing. Maybe what Lorena needed, then, was a way to learn facial expressions in a simpler form first, then build from there.

Soon after that, Kayla presented Lorena with *The Story of Little*

Bear, a forty-page computerized picture book that followed a baby bear on its adventures. On every page, Lorena learned a new noun or verb, like *honey* or *eat.* After a few pages, once Little Bear had eaten his fill, a bright, three-part chord burst from the computer, followed by a picture of Little Bear smiling. A new type of word materialized before Lorena's eyes: *happy.* "See? Little Bear is happy," Kayla explained. "I'm happy, too." Kayla smiled wide. "Let's both be happy." Kayla held up a hand mirror to Lorena. "Can you show me happy?"

At first, Lorena's poker face didn't budge. Or occasionally, when Kayla smiled, Lorena frowned, or when Kayla frowned, Lorena smiled, prompting Kayla to wonder *Is Lorena just messing with me?* But as the weeks went by and they continued practicing, Kayla could swear she saw the edges of Lorena's lips curl upward on command. As they worked their way through *Little Bear,* Lorena learned all six universal emotions: *happiness, sadness, anger, fear, surprise,* and *disgust.* Each was paired with an appropriate plot twist, picture, and three-part harmony that conveyed the spirit of the emotion. *Anger,* for example, was conveyed with a low rumbling of notes, *fear* with a tinkling of high ones.

As weeks passed, the sounds, animations, and practicing in front of the mirror started to sink in. Lorena's face came alive. She started smiling when happy, frowning when sad, and recognizing these facial cues in others. Lorena even learned a few facial expressions that weren't part of the basic repertoire, like the Mona Lisa. Lorena, with her oval face and dark wavy hair, bore a striking resemblance to the *Mona Lisa,* according to Kayla's dad, Rory. "How's my Mona Lisa?" Rory would say during family gatherings. Since then, this question had evolved into "Hey Lorena—do the Mona Lisa for me!" In an instant, Lorena would strike the pose: eyelids aflutter, a small smile on her lips, as if she knew the best secret ever, but would never tell.

Soon, at Lorena's urging, every room in the Picones' home had a small, framed print of the *Mona Lisa*. Lorena even carried a picture of the famous painting in her pocket, since gazing at it calmed her and gave her confidence. Now that she was fluent in the language of facial expressions, Lorena's ability to articulate her feelings blossomed in other ways as well. One day, when her mom, Alisa, was running the blender, Lorena, rather than screaming with her hands over her ears, said simply, "Turn it off, it hurts," before walking away. At the Cornales' massive family gatherings, Lorena, once unable to clear her plate herself, was now not only taking her dishes to the sink on a regular basis, but correcting other people's manners as well. "Wait," she once said before the family dug in to dinner. "We have to say grace first."

Occasionally, Lorena's social skills appeared to ascend to empathic levels unheard of in kids with autism. When Kayla showed up on crutches due to a basketball injury, Lorena was beside herself with worry. "What's wrong with your foot?" Lorena asked. "Does it hurt?" Lorena was so distressed, it was almost as if she could feel Kayla's pain. Perhaps she could.

As results rolled in from Lorena and the other kids in Kayla's pilot program, Kayla continued entering science fairs. By the time she graduated high school in 2007, she'd racked up more than $15,000 in prizes and scholarships at the Intel International Science and Engineering Fair and other competitions. But what truly bowled Kayla over were the crowds of parents who came up to her, often in tears, to say they had kids with autism and that they hoped to try Kayla's program themselves. These parents desperately wanted to know their children, but autism was getting in the way. Kayla knew how they felt and decided to help.

As science fair prize money rolled in, Kayla used those funds to apply for a patent and roll out her program in special education classes in more than twenty schools across Canada and the United States.

By her senior year, Kayla had grown, slowly but steadily, to a statuesque five-nine. She was now a starting point guard and captain of her high school basketball team. At school, she was poised at the top of the popularity pyramid and could have had her pick of parties, pals, and boyfriends. But more often than not, Kayla preferred to just hang with her buddy Lorena. In spite of the age difference between them—by this point, Kayla was seventeen, Lorena ten—the two had gotten so in tune with each other's life that they gossiped like old girlfriends. Once, during a visit, Kayla took a look at Lorena's mass of black curly hair, which was the exact opposite of her straight blonde tresses. Maybe that's why Lorena was so fascinated by her hair and insisted on fussing with it so often. Suddenly, it occurred to Kayla that she'd like to return the favor.

"Hey Lorena—how about we iron your hair straight?"

Lorena, with her cherub face and *Mona Lisa* smile, was already striking for a ten-year-old. With straight hair, however, she looked magnificent. Amid whistling from family members and clicking cameras, Lorena struck her signature pose, along with her small secret smile. Lorena was growing up. It was one of those moments that would stick with Kayla long after she was forced to tell Lorena the hardest thing she's ever had to say.

Kayla was leaving. She was graduating high school and going to college at Stanford. To send her off in style, Kayla's parents threw her a going-away party. Yet again, Kayla was surrounded by the Cornale clan and her thirteen cousins, who offered toast after toast about how much she'd be missed. The only family member who didn't put her feelings into words that night was Lorena. Perhaps she wasn't fully aware of what was happening. Or perhaps she was, and too sad or mad to talk about it. In spite of all the Sundays Kayla had spent one-on-one with Lorena, it was still hard to say for sure what thoughts were tumbling around inside her cousin's head, round and round, never to see the light of day.

Kayla did her best to soften the blow. "I'll see you again in a month or two, okay, Lorena?" she said cheerily at the end of the night. To Kayla, two months seemed fairly soon. But what did that mean to Lorena? While Lorena had learned to express many things, she still wasn't effusive when it came to affection. Hugs and kisses made her squirm. Bold declarations like *I love you* weren't part of her repertoire. Even though Kayla told Lorena fairly frequently that she loved her, Lorena had never uttered those words back. Kayla had come to accept this and not take it personally.

That year, come the second Monday in October, Kayla spent Canadian Thanksgiving alone. American colleges didn't allow time off for Canadian holidays, and besides, even if they had, it was a long flight for a three-day weekend. All day, Kayla's cell phone rang from family members saying that they wished she were back in Dundas. At one point, the phone was passed to Lorena, who, clearly annoyed by all the adults eavesdropping, ran off, holding the phone hostage.

"Uncle Rory's being weird" was Lorena's first complaint to Kayla. Translation: *Can't a gal have some privacy?* This was followed by a far bigger complaint. "I miss you." Translation: *Why did you leave? Are you ever coming back? Have you forgotten about your old pal Lorena?* Kayla told Lorena that she missed her, too, and would fly home as soon as she could.

At Stanford, Kayla quickly found that the major she'd set her sights on, biology, wasn't as intriguing as she'd hoped. Instead, she found herself drawn to a linguistics class where she worked with a roomful of nursery school kids, analyzing how they learned language. Kayla was fascinated to learn that at age three, kids understood the word *more* but not *less*. She started taking a sign language class with Cathy Haas, a professor whose claim to fame was that she'd taught sign language to Koko the Gorilla. Kayla was surprised to learn that Professor Haas had been deaf since age three due to scarlet fever, but neither her parents nor her teachers realized this until she was eight years old. They just assumed she was learning disabled, and meted out punishment whenever she pronounced a word wrong, which was often. Since then, deaf awareness had come a long way. Today, many didn't even consider being deaf a disability—just different. Kayla couldn't help thinking that perhaps one day, autism wouldn't be seen as a disability, either. Just different.

One day at the airport, Kayla struck up a conversation, in sign language, with a deaf girl about her age as they waited for their flights. Although their exchange was about nothing special, Kayla was amazed by the fact that had she never bothered to learn sign language, she never would have gotten to know this person. Next, Kayla started taking Spanish. The day she tried out a few phrases on her college cafeteria's Mexican chef, he was impressed. Soon after that, Kayla switched her plans from majoring in biology to linguistics. Language, Kayla realized, was her true love. Without it, she never would have gotten to know her cafeteria chef or that woman in the airport. Or Lorena, for that matter.

Over time, at Stanford, Kayla forged relationships with a new group of friends who warmly kidded her about her Canadian quirks, such as saying *aboot* versus *about,* or using the word *toque* versus *hat.* Soon, Stanford, which once seemed like an alien planet, felt like home. Back in Canada, Lorena Picone was adjusting to some changes of her own. Thanks to her development of the Sounds into Syllables program, Kayla had been selected as CNN's 2007 Hero of the Year. Eager to film footage of the little cousin who had inspired her program, camera crews flooded the Picone residence, as Lorena's parents braced themselves for disaster. Normally, all the bright lights and fast-talking strangers would have proved far too overwhelming for Lorena to handle. Yet, to their surprise, Lorena was comfortable with and captivated by the cameras, as well as the film director, whose hair—long, straight, jet black, with purple highlights—fascinated Lorena, much as Kayla's had years earlier.

Weeks later, Kayla flew to New York for CNN's award ceremony. Anderson Cooper served as host for the event, with musical interludes by Mary J. Blige and Sheryl Crow (no Shania Twain, alas). Harry Connick Jr. presented Kayla with her award, noting with a wink that he hoped it was *his* songs that had inspired Kayla and Lorena to embark on this project years earlier.

That night, Kayla received an award of $35,000, but for Kayla, the most rewarding part of that evening wasn't the money, but something else. Before the awards ceremony, CNN's purple-haired film director had informed Kayla's mom that they had a surprise in store for Kayla, and that it would become instantly clear what that surprise was once Kayla saw the video footage they'd taken in Dundas. So far, CNN had kept these videos under wraps, to air live that night during the awards ceremony, as well as on TV. Up in Dundas, the Picones were also watching, and waiting. Once Lorena saw herself on TV, she blushed, and murmured, "You remember that!" to herself, but otherwise she remained calm, as if she knew what was about to happen.

As Kayla stared up at the twenty-foot-high screen on stage, she saw Lorena, sitting at the piano where it had all begun four years earlier. Kayla could still remember the awkward silence that had settled over them like a fog on their first Sunday together, when Lorena refused to touch the piano keys and Kayla groped for a way to reach out to her little cousin. Now, seeing her up on screen, sitting there calmly in front of the camera, taking cues from a film crew, it occurred to Kayla that Lorena had come a long way. But what Lorena did next floored Kayla far more. Lorena turned away from the keyboard, looked straight into the camera, and said a small smattering of words that, up until this point, Kayla had never heard her cousin say.

"I love you, Kayla."

Kayla knew that Lorena loved her. But hearing it that day, at the CNN awards ceremony, was different. For that one moment, even though they were thousands of miles away from each other, Lorena seemed to be speaking directly to Kayla, and speaking from the heart.

"I think there are tons of people who truly want to help children with autism," Kayla said. "But I think many people have this defined model in their minds of how learning should take place. We need to detach ourselves from these predetermined notions, slow down, and take some time to get to know these kids. Their idiosyncrasies are the coolest part. I've definitely learned to be more open-minded."

While this wisdom may be especially true for kids with

autism, it struck me that it's true for just about everyone we meet. Maybe we could all stand to listen a little more closely to the people around us, whether they're friends, family, cousins, or even six-year-old kids. Ask a question, and you never know what they'll say.

THE NEXT BILL GATES

*A man is rich in proportion to the number of things
which he can afford to let alone.*

—HENRY DAVID THOREAU

"**B**ut did any science fair kid strike it rich?"

Whenever I posed this question to attendants of the Intel International Science and Engineering Fair 2009, the answer was unanimous: I had to meet Philip Streich. This eighteen-year-old's status was so legendary, students pointed and whispered when he walked by. Girls sighed and asked their friends, *Is he single?*, since in addition to being smart, Philip was cute—think Tom Cruise from *Risky Business* and you get the idea. Philip even had an entourage, a tight-knit posse of science fair friends who flanked him wherever he went. When not talking to them, he typically had a cell phone glued to his ear. Fighting my way through these layers of minions and admirers would require some paparazzi-level skills on my part.

"Five minutes," I begged Philip at one point when he was in between phone calls.

"Oh shoot," Philip said apologetically. "Can we talk to-morrow?"

"Just one question," I said. "How do you think you'll do this year?"

Philip blushed. One of his henchmen spoke for him. "Philip will place first." Of course. Then they were gone.

Philip's trek to science fair fame began two years earlier when he swept through Intel ISEF 2007 and racked up eight awards. Over a celebratory dinner that evening, when someone at the table whipped out a calculator and summed up his earnings, toasts ensued over the total: $66,150. At Intel ISEF 2008, Philip racked up a slew of new awards, and Intel ISEF 2009 would no doubt be no different. Prize money wasn't the only cash lining his pockets. Philip's project—which found a way to cheaply and simply mass-produce graphene, a substance hailed as a faster, cheaper replacement for silicone—had since been transformed into a company called Graphene Solutions, which was predicted to generate sales of $12 million in the first five years.

Philip, I assumed, must have been groomed to succeed, surrounded by every opportunity a young, smart mover and shaker should have. Up until the age of ten, this was true. But September 11, 2001, had a way of changing things.

Philip grew up in Princeton, New Jersey. His parents, Amanda and Joel, met at Harvard Business School. After graduating, they both got jobs at J.P. Morgan, where they would rise through the ranks to become vice presidents. Once they decided to have kids, they tackled this task with the same alacrity as they had their careers, giving birth to three—first Philip, then Peter, then Caroline—in three and a half years. At this point, Amanda quit her job and poured all her energy into her new role as power mom. She shuttled Philip to piano lessons, Peter to Japanese class,

Caroline to tennis, and whatever else caught their fancy. Amanda's full-throttle child rearing quickly produced results. At age six, Philip sat down at the piano and played Bach's Toccata and Fugue by ear, convincing hordes of parents to sign up their own kids with his piano teacher. At school, Philip earned rave reviews from his teachers. He was the envy of every achievement-oriented überparent in town.

Philip thrived on the attention his talents brought him. During his monthly piano recitals, Amanda found it a little disconcerting that her son never once looked at the keys and instead stared straight at the audience. Since Philip's enthusiasm was contagious, he had little trouble convincing others to go along with whatever plans he cooked up. In kindergarten, he gathered his classmates on the playground and taught them how to sing the "Hallelujah Chorus," with him as their conductor. In third grade, Philip's productions became even more elaborate, with a Blues Brothers concert featuring homemade costumes, coordinated dance moves, and himself on the harmonica. While Philip had plenty of friends, he was especially close to one boy named Danny Dunn.* Together, they built robots. When Philip got his first chemistry set, the twosome sequestered themselves in the basement, mixing chemicals, utterly convinced that their efforts would spontaneously generate life.

Since Philip and Danny were friends, the Streichs were also acquaintanced with Danny's parents, a power couple much like them, named Don and Meredith. Don worked in finance just like Joel. Their offices were just a few blocks from each other in downtown Manhattan. Meredith and Amanda got to know each

* The names of Danny, Don, and Meredith Dunn were changed since they could not be reached for comment.

other chaperoning their two boys on school field trips. Once, during a trip to Ellis Island, the four of them waved toward the World Trade Center towers, where Don worked on the top three floors, at Cantor Fitzgerald. Meredith had taken out her cell phone and called Don, who'd grabbed a pair of binoculars and claimed he could see them, waving, even though he was miles away and a hundred floors up. Amanda remembered hearing Danny yell, "Hey, Dad, look at me!" Six months later, this memory would haunt Amanda as she picked up the phone to hear her husband Joel's voice.

"Have you heard what's going on?" Joel asked. "Turn on the TV."

It was the morning of September 11, 2001. Amanda watched as one tower of the World Trade Center collapsed, then the other. Over the years they'd worked in the finance industry, Amanda and Joel had gotten to know so many people in those towers. Terrified, Amanda spent the rest of the day fielding phone calls from people asking if Joel was okay, as Amanda tried to check in with Joel on his cell phone with no luck. Once Philip and his siblings arrived home from school, Amanda gathered them together and waited. Finally, at some point late that night, Joel, through a combination of boats, buses, and walking, made it home. He was covered in white dust—the ashes of what was left of the World Trade Center. Amanda hugged him anyway. At least he was safe. But the same could not be said for others.

Don Dunn's body was never found. His family was forced to presume he went down in the rubble. Nonetheless, for a while, Danny trained a magnifying glass on photos of victims who'd jumped, straining to see if his dad was among them. Any answers, even bad answers, would have been better than nothing.

Days later, the Streichs attended Don's funeral—the first of many they would attend in the coming weeks. Philip, who could typically put a smile on anyone's face, especially Danny's, was at a loss as to how he could make things better. Then Amanda and Joel sat their kids down and made an announcement. Joel had quit his job at J.P. Morgan. They were moving away from Princeton, away from their friends and family, away from 9/11. After twelve years in this rat race, they had had enough. They were going to ditch the life they knew for good.

Philip was stunned. Leave home? Leave his friends? And where would they go? Joel and Amanda owned some farmland in Platteville, Wisconsin. On this land, there was a house that had been abandoned for a while, but with a bit of elbow grease, it could be fixed up. The farm was in Amish country, which might be a nice change of pace. But Philip wasn't buying it. Farm? Fixer-upper? *Amish country?* Had his parents gone insane? Philip prayed it was just a phase as they reeled in the wake of 9/11. But by that summer, the Streichs had packed their belongings and headed west. As the scenery outside the car window flashed by, Philip watched as the brick mansions and bustling streets he was used to grew more and more sparse, until finally, all he could see, stretching all the way toward the horizon, was fields of corn.

Philip was moving to the middle of Boring Nowhere. He couldn't imagine feeling more miserable. Then he got a look at the house he'd be forced to call home.

To add some levity to their situation, the Streichs would jokingly refer to their home as the Amish Crackhouse. They'd been warned the house was old, more than a hundred years old. Amanda, who was sick of the McMansions back east, had considered this a good thing at first. But as their car pulled up the quarter-mile-long driveway, the reality dawned on her that *old* meant more than a house with character. Most of the windows were broken. The front porch was so run-down, they feared it would collapse as they crept across the creaky floorboards. The front door was so warped it had to be kicked in rather than opened. Inside, the linoleum flooring was peeling, the carpet covered in a layer of fur from some sort of animal, maybe cats, although it was hard to say for sure. The walls had little insulation to speak of, which would make winters miserably cold. The pipes inside the walls were rusty and would need to be replaced. In addition to these squalid conditions, the house was so small that the family of five would be forced to share one bathroom. Philip, who'd grown up with his own bedroom, would now have to share a room with his younger brother.

That first night, as the Streichs tried to get to sleep surrounded by boxes, Amanda lay in bed and stared at the ceiling, haunted by one question: *What have I done to my family?* She prided herself on being a risk taker. But this time, she feared she'd gone too far. Before leaving Princeton, they had been to many wonderful going away parties. Friends and family expressed shocked delight, tinged with a streak of skepticism, at the fact that the Streichs were opting out of the rat race. "I've always dreamed of moving to a farm," one guest said and sighed dreamily. But the Streichs were also a cautionary tale. Behind their backs, a few undoubtedly whispered that 9/11 had pushed them off their rockers. Families might *fantasize* about moving away from it all, but doing it? Well, that was just nuts. Sooner or later, they'd regret this impulsive decision. They'd be back. Lying in bed that night, Amanda wondered if the cynics were right.

The next morning, as the sun peeked through broken windows, the Streichs explored their property, which was in even worse shape than the house itself. The pond and yard out back were so littered with beer cans they came to suspect that their home had been turned into some sort of party house since its desertion. Many of the surrounding sheds had been taken over by hissing raccoons. Philip and his siblings attempted to turn one shed into a clubhouse before it collapsed, thankfully when they weren't inside. Every day there was a new disaster, and the repairs kept Philip so busy he didn't have time to feel sorry for himself. Every hour of the day was filled with nailing, prying, or carting junk somewhere in an effort to make the Amish Crackhouse a home.

The work level ramped up another notch once the animals arrived, starting with a herd of cows. While Amanda assumed the cattle couldn't jump the gates they'd erected to keep them cordoned off to certain areas, one day she was startled to see a cow standing in front of her newly renovated front porch, all but ringing her doorbell. Flocks of sheep seemed more manageable, and Philip's sister, Caroline, grew attached to the docile creatures and would bottle-feed a few who seemed in need of extra care. Then one day, one particularly precocious lamb, nicknamed 40F due to its ear tag, wrenched its neck while stuck in a feeding bin and died. Caroline was horrified. At another point, their black Labrador, Lily, got lost in their soybean fields, which confounded the dog's sense of smell and direction. Search teams found Lily days later, lying in the field dead, most likely due to heatstroke and dehydration. Now, all they had left of Lily was an urn with her ashes and her paw prints in a piece of wet cement they'd poured during the first wave of renovations.

What could go wrong with chickens? The Streichs figured

they'd be a safe bet. But no one had warned them about the rooster, which came tearing at Philip one morning while he collected eggs. Philip fled and burst breathless into the house, where his mother, Amanda, was still in her pajamas, sleepy and slightly amused that a rooster could frighten her son so much. To demonstrate what he was dealing with, Philip put on his father's boots, which were thick enough to protect his legs, then walked back out as Amanda watched from the window. The rooster, seeing Philip, shrieked and tore at him with vicious intensity. Amanda's jaw dropped as Philip ran back inside.

Joel was away on a trip. Amanda couldn't call her neighbors, since she didn't really know them. Out here in farm country, Amanda had wondered whether her neighbors might swing by with apple pies, smiles, and gossip about goings on in town. Now she realized how silly this notion was. Her family was alone and on its own. They were hostages in their own home.

That's when Philip said something Amanda never thought she'd hear her son say.

"Can I shoot it?"

Amanda had never cared for guns. But out here, she could see how a gun could come in handy. Joel hunted on occasion and owned a rifle. He had also taught Philip how to use it. Amanda really, really didn't want her son shooting anything. But since they couldn't leave their home and Joel wouldn't be home for days, what choice did they have?

"Okay," Amanda said. "Just be careful . . ."

Philip stepped out onto the front porch, rifle in hand. His mother trailed behind him, wielding a broom as added protection. The rooster, seeing them, screamed and hurtled toward them, as Philip raised the rifled, aimed, and fired. Philip hadn't shot guns much. It just wasn't something kids did back in Princeton, New

Jersey. But out here, if he were to survive, he'd have to learn a whole new set of skills.

Philip got the rooster on his very first shot. It fell, spasmed a few times, then lay still, blood soaking the grass beneath it. Philip and Amanda cheered. They'd won their first battle. Only once Joel returned home, a squabble ensued.

"Why'd you shoot it?" Joel asked Amanda. "Don't you think you were overreacting?"

"If you were here, you wouldn't have said that," Amanda replied. She hated when Joel left on long trips. She could swear the animals could sense when her husband wasn't around and chose those moments to act up. Another time Joel was away, a cluster of cows decided to give birth right in the middle of a torrential two-day thunderstorm. Mixed in with the constant cracks of lightning and howling wind, Amanda and her kids could hear the cows bellowing as they gave birth to their calves, which were so small that they slipped out under the electric fence and tottered off into the cornfields. Unless someone retrieved the calves and placed them back over the fence with their mothers, the calves would die.

Philip knew what he had to do. He and Amanda crept out of the basement and into the fields, where the cornstalks clattered like bones. One by one, Philip found the bleating calves and hauled them back over the fence. The mother cows, unhappy with the fact that Philip was handling their offspring, threatened to charge. For the hundredth time, Amanda regretted moving to this inhospitable place.

Yet within that two-day storm was a silver lining. Philip had changed, in ways that a posh enclave like Princeton, New Jersey, could never have changed him. Trial by trial, he was learning that if he didn't deal with the crisis in front of him, no one would.

Out in this Amish Crackhouse in the middle of Boring Nowhere, Philip was growing up.

In between the renovations and animal rescue missions, the Streichs also struggled with the question of what to do about school. Out here, the school systems were no more challenging for Philip than they'd been back in Princeton. In seventh grade, Amanda started home-schooling Philip, with plans to transition him into a high school later. Only where could he go that would keep him engaged?

Still longing to return to his old life back east, Philip applied to boarding schools like Andover in Massachusetts and Exeter in New Hampshire. Once the acceptance letters rolled in, the Streichs took a tour of each campus and couldn't help but be impressed. At Exeter, they stared agog at a magnificent skeleton of a whale, which was more than most museums had at their disposal. Still, something wasn't quite right. In class, the kids seemed slightly bored. Plus, hovering like a fog among the pillared buildings and well-modulated lectures was an ambience that felt strangely familiar. It felt a lot like life back in Princeton.

Back when Philip was in primary school and Amanda was casting about for ideas on how to challenge him intellectually, one Princeton übermom had suggested that Amanda send her son to an after-school program called Kumon. Kumon, after all, was the best. Curious, Amanda went to Kumon to check it out. She saw rows and rows of kids filling out booklets. Candy was awarded to kids who filled out their booklets correctly. Amanda cringed at the thought of sending her son to such a place. It seemed so Stepford-esque. Right before the Streichs left Princeton in the wake of 9/11, an überdad whose daughter was in the same grade as Philip had made a confession.

"Part of me is glad you're moving," he admitted to the Streichs, half joking. "Philip's tough competition. Now my daughter has a shot at being top of the class."

Amanda knew that Philip was unusual, and that as a mom, she was lucky. At seventeen, he would score a perfect 2,400 on his SATs, no prep course required. Amanda could also understand parents' desire to do everything in their power to encourage their kids to excel. Still, in her heart, she had a hunch that filling out booklets for candy at Kumon wasn't the answer, whether a child's intelligence was average or off the charts. Kids needed to do things because they *wanted* to. Without that, they would eventually lose that spark that jump-starts all their endeavors. She didn't want to see that happen to Philip.

At this point, their options were dwindling. Philip wouldn't thrive in a public school in Wisconsin. Now it was clear that private school back east wasn't quite right, either. So where should Philip go to school? Where, exactly, would he fit in?

That's when it occurred to Amanda that the Amish Crackhouse in the middle of Boring Nowhere, for all its flaws, had served as quite a learning experience for her son so far. Maybe homeschooling—which Amanda had considered a stopgap measure—could also work once Philip reached high school. Maybe she had more to teach her son than she thought.

"Let us pray."

Amanda was seated at a home-school association meeting in Platteville, Wisconsin. Many of her neighbors taught at home for religious reasons, so she had expected to encounter a bit of religious zeal. What she hadn't expected were the politics.

"Dear Lord, we pray that George W. Bush will be reelected President . . ."

Amanda froze. No way was she praying for that. She and Joel were staunch Democrats; they even had a donkey named Demmy. In Princeton, their political views meshed seamlessly with their neighbors'. Out here, in this home-school meeting, they weren't merely the minority. They could very well have been the *only* Democrats in the room. Amanda decided it was probably best to keep her political opinions to herself. In fact, it was probably best for the Streichs to keep many of their opinions to themselves. During another home-school meeting Amanda attended with Philip, students were asked to give a speech about their favorite book. Philip, who wasn't shy, was one of the first to stand up and talk about how much he loved the *Harry Potter* books. A hush fell over the kids.

"That's nice," the teacher sniffed. "But I prefer books that have a *moral* message." The next three kids who stood up said the Bible was their favorite book.

Soon after that, Amanda scaled back on attending home-school meetings, deciding that she and Philip might be happier doing their own thing. While Philip's sister, Caroline, remained happily in public school, his younger brother, Peter, quickly joined the home-school contingent. Peter's interest in foreign languages had attracted some flack from public school teachers, who grew alarmed when Peter showed up at school with books in Arabic. Every morning, Philip and Peter woke up and came downstairs to the dining room, which they'd established as home-school central. Then they'd stare expectantly at their mom, as if to say: *What now?*

At first, Amanda wasn't sure. But having volunteered as a teacher's aide on occasion, she did know a few things. For one, she knew that a lot of time in school was spent just keeping kids busy. Only busywork didn't challenge kids. So instead, she tried a different tactic.

"Why don't you boys just do whatever you want?"

It was a risky move. It could easily backfire. Given such freedom, what would they do? Become addicted to daytime TV? It was possible, although Amanda was banking on the hope that her kids could come up with something better than that. After all, this wasn't just a weeklong vacation. They were free indefinitely. It was exhilarating, but also frightening. Philip was forced to ask himself: *What do you want to do—really?*

Philip picked up a book about mathematics and started flipping through it on the front porch. Peter grabbed a book about foreign languages and retired to the kitchen. The house was quiet. At first, Amanda wondered if they were bored. But maybe kids actually *needed* time to be bored. Only then could they find their own bearings. Now that learning was no longer imposed on them, they could finally *choose* to learn and let their thoughts run wild.

As the days passed, Philip zoomed through book after book. After devouring a book on nuclear physics, the sun morphed before his eyes into millions of hydrogen atoms fusing together. After reading a book on electricity, his cell phone turned into a symphony of intricately orchestrated circuits. As Philip wandered around the environs of his mind and pushed the limits of his imagination, Amanda pored over home-schooling books and websites. She learned that while most families home-schooled for religious reasons, a small but growing minority did so because they were disillusioned with America's education system. Abraham Lincoln, Albert Einstein, Franklin Delano Roosevelt, and even John Witherspoon—the president of Princeton University—were home-schooled. So the Streichs were in decent company.

Days, weeks, then months passed as Amanda, Philip, and Peter acclimated to their own surreal version of school. Soon, Amanda noticed that her unorthodox teaching methods had made her kids different. When friends from around town came over to play, they couldn't deal with not having something specific and structured to *do*. Immediately upon setting foot in their home, the kids would ask if the Streichs had video games or, barring that, if they could watch TV. Only Amanda's kids weren't interested in either of these pastimes. "How about a board game?" Amanda suggested at one point. One boy was lukewarm about that, but agreed to play a couple rounds of a card game called Phase 10. Soon, the boy was having so much fun, he wanted to keep playing and playing.

While Philip's interests ranged far and wide, he began noticing that one subject in particular seemed to embody all of the elements he loved most: science. With science, he could get his hands dirty and build things. He could apply what he'd learned in a real-world setting. He could make a palpable difference in someone's life, even his own. Amanda suggested Philip try taking a few college courses at the University of Wisconsin in Platteville. There, Philip would meet a professor who would pick up where Amanda left off and steer Philip's unbridled love of learning out of the Amish Crackhouse and into a lab.

"So do you really think that carbon nanotubes might be soluble?"

James Hamilton and Jonathan Coleman were sitting in a pub in Dublin, drinking pints of Guinness. For Jim and Johnny, who were professors in the field of nanotechnology, discussing the solubility of carbon nanotubes was the equivalent of pondering whether the Red Sox would ever win the World Series. Through the years, they'd examined this problem from countless angles.

Most nanotechnologists had long ago given up hope. But today, in a pub, over pints of Guinness, Jim could see a light at the end of the nanotube tunnel.

Nanotechnology is the study of matter on an extremely small scale. Nanoparticles are about 1/100,000th the size of a hair, just a tad bigger than a molecule. Nanoparticles could be pieced together without the defects that typically plague larger objects, and that's why nanotechnologists had been eyeing carbon nanotubes with particular interest. Carbon nanotubes, which are atoms of carbon linked like chicken wire rolled into a cylindrical, drinking straw–like structure, were five hundred times stronger than steel and one thousand times more conductive than copper. That made carbon nanotubes a potential cash cow with plenty of practical applications, from paper-thin TV screens to bridges with cables the thickness of string. The only problem was that carbon tended to clump together, which rendered the nanotubes about as commercially useless as a pile of soot, which was exactly what carbon nanotubes looked like. Nanotechnologists had tried to unclump carbon nanotubes using every substance under the sun—alcohols, oils, paints, detergents, even salmon sperm DNA. And yet, five years and four thousand plus academic papers later, nanotechnologists were still scratching their heads.

Jim perked up when Johnny mentioned he'd made progress mixing carbon nanotubes with a solvent called N-Methylpyrrolidone, or NMP. While it seemed like the solvent was doing *something,* Johnny had no precise way of proving it. Without proof, he couldn't publish or take credit for his accomplishments. Jim got to wondering if he could help.

Jim had already made a name for himself in the field of nanotechnology. Years before, he'd created a polymer that was so effective at removing dust and debris from uncleanable surfaces, he was contracted to clean the optics for the Hubble Space Telescope and the Hope Diamond. To clean the jewel, Jim got to enter the vault in the Smithsonian's National Museum of Natural History and hold the diamond in his hand for a full half hour. It was ice cold and quickly drew the heat from his hand. The diamond was beautiful, and Jim's polymer made every facet even more brilliant. It was moments like these that made being a professor on the small, snowy campus of the University of Wisconsin in Platteville totally worth it. Jim, unlike most professors, saw little use in sequestering his research in an ivory tower. He preferred to get his findings out into the world and put them to practical use.

In addition to Johnny's promising work with the solvent NMP, Jim had another reason to be hopeful on the carbon nanotube front. Recently a research assistant named Philip Streich had joined his team. While only fourteen, Philip struck Jim as amazingly bright. Bumping into each other on campus, the twosome had ended up chatting about a variety of topics—chemistry, literature, philosophy, opera, ethnic food, on and on it went. Jim quickly realized that there were few things Philip *didn't* relish talking about and that the two of them even shared the same tastes in food. Later on, whenever they'd eat lunch together, they'd inevitably pick the same thing off the menu. Jim had heard good things from a fellow professor about Philip's research skills. So when Philip asked Jim if there was room in his lab for him, Jim happily gave him a tour.

The minute Philip set foot in Jim's lab, he was fascinated. Huge, humming instruments packed the counters. Jim explained that they were lasers—very powerful lasers. One, if pointed at the wall, could burn the paint off in an instant or, with repeated zaps, could blast a craterlike divot. Certain lasers were kept in windowless rooms and could put on a light show worthy of a Pink Floyd concert. Philip's first task in this lab would be to take a solvent called NMP, add carbon nanotubes, and see if he could determine if any nanotubes dissolved. Eyeballing a beakerful, of course, wouldn't provide the precise evidence they'd need. That's where the lasers might come in handy, providing measurements of light scattering through the liquid. Jim knew he was asking Philip to solve a problem that had stumped countless veteran researchers before him. But Jim also knew that a fresh set of eyes like Philip's could often see things that others could not. It was a long shot, but what did they have to lose?

First, Philip immersed himself in the literature, poring over the thousands of papers detailing attempts to unclump carbon nanotubes, all of which had failed. Eventually an article from 1946 on Debye light scattering caught his eye. Feeling hopeful, Philip tried this technique using every laser in Jim's arsenal, as well as lasers at other universities. But none of the equipment was sensitive enough to detect what was happening.

At this point, Philip could have done what every other professor, grad student, and researcher who had ever tackled this question had done: given up. No one would have thought less of him for leaving the blood, sweat, and tears of finding a solution, if a solution even existed, to someone else. But one thing Philip had learned during his years in the Amish Crackhouse was that giving up was not an option. Had Philip stayed in Princeton, he would have gotten used to leaning on others and knowing that somewhere, someone would pick up the slack. But out here, the answer was obvious. If an instrument didn't exist that was sensitive enough to do Debye light scattering on carbon nanotubes, Philip would have to build it himself.

. . .

Philip pieced it together, part by part, using spare components lying around in the lab. "Wow" was all Jim could say at first. He named it the Streichometer. But did it work? To find out, they aimed the laser into batch after batch of carbon nanotubes swimming in NMP solvent. After collecting the data, Jim was stunned to see that Philip's light scattering instrument didn't just work, it worked better than anything on the market. After more rounds of tests and trials, Philip, Jim, and Johnny in Dublin published their results. It was official: Carbon nanotubes, in spite of all expectations to the contrary, were soluble. Out of these tiny, strawlike nanostructures, a small team of researchers, including a fourteen-year-old boy, had hit upon something really, really big.

During science fair season his sophomore year, Philip presented his work on carbon nanotubes at Wisconsin's state science fair. There, he sailed to first place, which qualified him to present his research at the Intel International Science and Engineering Fair 2007 in Albuquerque, New Mexico. Philip's mom, as his teacher, attended with him, figuring it might be fun. At the awards ceremony, she heard her son's name being called again and again. Science fairs, she realized, weren't just some quaint pastime for kids. They were the secret portal to Philip's salvation. Here, he was in his element. After years of isolation on a farm in Wisconsin, Philip had found a club where he not only belonged, but was king.

After Intel ISEF, Philip returned to Wisconsin eager to expand on his success. The next logical step was to try dissolving graphite to produce graphene. Graphene was essentially what you got when you unzipped a carbon nanotube and laid it flat. Since electrons traveled a hundred times faster in graphene than in silicone, graphene could be used to shrink supercomputers to the size of a fingernail or to create efficient inexpensive solar cells. Back in the lab, Philip and Jim got to work. Soon, Philip's Streichometer struck gold again, enabling them to not only measure the solubility of graphene, but devise a way to mass-produce it. Jim, realizing that Philip's Streichometer had just spit out the equivalent of a winning lottery ticket, started filing for patents. Typically, professors grant a ninety-ten or eighty-twenty split with their students on patents. Jim split his with Philip fifty-fifty.

Two years, five patents, and more than $100,000 in science fair winnings later, Philip and Jim were having lunch at one of their favorite restaurants, Fiesta Cancun. As usual, they were eating the same thing, chili verde. In a few months, Philip would fly to Cambridge, Massachusetts, to start his freshman year at Harvard. Soon, his involvement in science fairs would come to an end. They had won him some sweet scholarships and co-ownership in five patents. At this point, Philip had outgrown science fairs, much like kids outgrow amusement parks. Only he wasn't ready to get off the ride just yet.

"So about this company, Graphene Solutions," Philip ventured, referring to the company Jim had launched to turn their patents into a profitable enterprise. "How about we co-own the company fifty-fifty?"

Jim looked up from his chile verde in disbelief. There was no way Philip would have time to do that with his course work at Harvard. But Jim could see that Philip meant what he said. He didn't want to be a typical Harvard grad, going to Finals Club parties and picnics at Martha's Vineyard. He wanted to build solar cells and supercomputers the size of a fingernail. Running a company together would be tricky given the distance between them, but it wasn't impossible. When you found the right team, anything was possible.

Jim put down his fork. "I don't know about fifty-fifty," he said with a smirk. "How about seventy-thirty?"

"Fifty-fifty."

After more mock haggling, they shook hands. Fifty-fifty it would be. As Jim and Philip met with investors and entrepreneurs to gauge their interest in being involved in the enterprise, many marveled at how accomplished and polished Philip was for his age—and having been home-schooled on a farm, no less. But Philip had learned that wherever you live, you can find people who can keep life interesting. Growing up in the Amish Crackhouse in the middle of Boring Nowhere hadn't been boring at all. Not in the slightest.

Back in Princeton, New Jersey, as news of Philip's exploits traveled through the grapevine, leagues of überfamilies in their McMansions with kids in Kumon were undoubtedly ruffled. The Streichs were no longer a cautionary tale. Philip had gotten into Harvard, won more than $100,000 in scholarships, and co-owned a multimillion-dollar company. As strange as it seemed, the Streichs' harebrained scheme had panned out beautifully. Some überparents might have felt a twinge of jealousy, but many more who heard the tale were inspired. If the Streichs could do it, then maybe they could, too.

The Streichs weren't the only ones who'd drastically changed their lives in the wake of 9/11. Many people, after seeing the World Trade Center towers fall, felt compelled to look inward and ponder: *What do you want to do—really?* Some quit their Wall Street jobs and became watercolor painters or poets. Others moved off the grid, into fixer-uppers, or decided to have kids now rather than wait. Many more dreamed of veering off the beaten path but were too scared of the consequences. Still, if

9/11 taught us anything, it was that life is too short and unpredictable to play it safe.

This wisdom had gotten Philip far in life. Only how much farther would it take him at Intel ISEF 2009? That would depend, of course, on who he was up against.

THE SUPER BOWL OF SCIENCE FAIRS

You miss one hundred percent of the shots you don't take.
—WAYNE GRETZKY

R *rrrip.*

This is how the Intel International Science and Engineering Fair 2009 began, with the sound of box cutters slashing through duct tape. That's because the first task that all 1,502 competitors had to tackle once they arrived by plane, train, bus, or car was to head to the convention hall and set up their booths. Eager to sneak a peek at the projects as they emerged from their boxes, I flashed my press pass to the guards, entered the football field–sized Sparks Convention Center in Reno, Nevada, and trolled the aisles. At one booth, I gawked at a table tennis–playing robot. At another, I marveled over a soon-to-be-patented chemical spray that made dog poop disintegrate in mere days, thus eliminating the need to pick up after your pooch. As a dog owner, I couldn't wait to see this spray, which its inventor had dubbed "Dog Gone Poop," between the Swiffers and the Windex at Wal-Mart.

Intel ISEF is the world's largest international pre-college

science competition, representing the top fifteen hundred plus competitors from more than fifty countries. Projects span seventeen disciplines, from Chemistry to Computer Science, Engineering to Environmental Management, and each category has its own personality. Kids in Animal Sciences, for example, are softhearted souls in flannel and flip-flops. Kids in Medicine and Health are type-A pre-med perfectionists. Everyone pooh-poohs the poor kids in Behavioral Science for not being "real" scientists, while no one pokes fun at Physics and Astronomy, which boasts arguably the biggest bragging rights of all.

"We're the manly category," explained one physics veteran, puffing up his chest. "After all, we know how to blow stuff up!"

Intel ISEF, of course, bars students from bringing anything into the convention hall that could be dangerous. But that doesn't stop people from eyeing certain projects suspiciously, and few raised more concerns than the nuclear fusion reactor that was wheeled in by Taylor Wilson, aka the "Radioactive Whiz Kid." As crowds swarmed his booth to inspect the gleaming machine from a safe distance, pint-sized Taylor politely fielded questions in his choirboy-high Southern drawl. At one point, late-night TV show host Conan O'Brien swung by with his camera crew, having hit upon the bright idea that an international science fair was fertile territory for one of his comic skits.

"So," Conan said to Taylor, "you're building a nuclear weapon?"

Taylor said no. Conan wasn't convinced.

"So when I go back to my hotel room," Conan continued, "how many arms am I gonna have?"

Taylor was used to these wisecracks. The fourteen-year-old Reno, Nevada, resident also knew that to the average observer, his nuclear fusion reactor looked ominous, and that he looked far too young to be wielding its powers. Still, Taylor's very existence was living proof of one of the most basic tenets of science: that looks could be deceiving. Taylor saw this first-hand while talking to a seventeen-year-old from California who'd invented a device that detected bisphenol A, or BPA, in plastic.

"Hey look at this," he said to Taylor, holding up his water bottle. "It says 'BPA free.' I wonder if it really is."

"Why don't we test it?" Taylor suggested.

Together, they deposited a water sample from the bottle inside the detector. Within seconds, its light flashed, indicating that the water contained traces of BPA. The bottle's labeling was a sham.

"That's really cool," Taylor said. "I think you're going to win."

"No way," said the seventeen-year-old, eyeing the crowds surrounding Taylor's booth. "I think *you're* going to win."

They would find out who was right soon enough.

As I made the rounds, it quickly became clear to me that many competitors whiled away the early days of Intel ISEF socializing—and scoping out cute prospects. And BB Blanchard, for one, wasn't going to let a little thing like leprosy stand in her way. For this sixteen-year-old from Baton Rouge, the flirtfest began the instant she and her partner, Caroline Hebert, arrived at their booth and opened their boxes. During shipping, the hinges to their display board had shaken loose, and they would need to be drilled back together. Only neither BB nor Caroline knew how to wield a drill. On cue, two boys from Alabama swooped in to save the day.

"Need some help?" the boys asked.

"Oh my god yes!" BB and Caroline sang in unison. "You guys are lifesavers."

After procuring a power drill, which fair administrators kept on hand for just such emergencies, the boys fixed the problem. The girls' board was ready to roll. And so were BB and Caroline, who wandered around the convention hall developing crushes left and right.

"One kid from Denmark had the cutest little accent," BB sighed to me as she recapped the highlights. Only how did students react when they heard she'd had leprosy? "Some people were like 'Oh my gosh that's so cool!' while others were like 'Oh, okay,' as if I'd mentioned I had diabetes," she said. "It was no big deal. No one freaked out at all."

The only negative reaction BB encountered was from a fifth grader who swung by her booth on a class field trip. "You have that? The Bible disease?" the girl asked, clearly horrified, but BB explained that having leprosy was nothing to be afraid of. In fact, in terms of her project, it could be seen as a benefit.

"You're going to win," one competitor told BB, pointing out that judges loved projects with a backstory. Perhaps this competitor was right. But BB wasn't banking on it. All around her, she saw projects that were so brilliant they blew her away. Win an award? Here?

No way.

After competitors had set up their booths and mingled, it was the perfect time to slot in some entertainment. Only what would a bunch of science fair kids be thrilled to see? While a U2 concert might impress your typical group of teenagers, for this group, the real rock stars worth booking were obvious: Nobel laureates.

This year's Intel ISEF attendants would get the privilege of meeting not just one or two Nobel laureates, but six, along with a handful of other world-renowned scientists. Students packed the auditorium, buzzing in anticipation, as intellectual heavyweight after intellectual heavyweight stepped on stage. Leon Lederman had won a Nobel for his work with neutrinos. Dudley Herschbach had made inroads toward understanding elementary chemical reactions. Kurt Wüthrich had developed nuclear magnetic resonance spectroscopy. As the panelists took their seats, the floor opened to questions. *Will we ever find a viable renewable energy source? What scientific topics are commonly overlooked? Do aliens exist?* While hearing Nobel laureates ponder these topics was exciting for everyone in the audience, for one student, it literally changed her life: the anti-nerd known as Eliza McNitt.

"I never considered myself a scientist. I considered myself an actress," Eliza explained. Art was her calling, her passion, her raison d'être. Lately, though, her research on Colony Collapse Disorder among honeybees had instilled in her a love of science. Still, the seventeen-year-old had always assumed that science and art were mutually exclusive endeavors. Sooner or later, she'd have to leave science behind. Right?

Not so, said the Nobel laureates. When they were asked, "Do you have any artistic talent, and has that helped in any of your scientific endeavors?" Douglas Osheroff, who'd discovered the superfluidity of helium-3, recited the poem "Reincarnation" by Wallace McRae. Martin Chalfie, who'd won the Nobel for his work on green fluorescent protein, admitted he played classical guitar. Meanwhile Jocelyn Bell Burnell, an astronomer whose work led to the discovery of pulsars, bent over her chair, rummaged through her purse, and pulled out a slim book. In this book, Bell

Burnell said, she'd collected poems about astronomy. She'd been collecting them all her life, but rarely talked about it. It was time, however, to come out of the closet. She loved science *and* art. It was the perfect marriage, in fact. "We need the rigorous testing, but we also need creativity, wild ideas that are off-the-wall," she said. "That's where hypotheses come from in the first place."

That day, Eliza learned something that she'd always somehow known was true. Science wasn't just some dry collection of statistics and theories, separate from art. Science *was* art. Science was poetry. Science was creative, exciting, even sexy. Two years ago, before giving science fairs a shot, Eliza would have sneered at such a notion. But now, like a prom queen who had deigned to date a nerd and realized that underneath his glasses he was totally hot, Eliza realized she was head over heels in love with science. She couldn't imagine life without it.

But did science love her back? So far, it hadn't exactly rolled out the red carpet. Back in Greenwich, Connecticut, she'd endured endless heckling from science nerds in her class. Here at Intel ISEF, she'd attracted strange glances for the way she dressed and danced. Like the honeybees that had served as the focus of her project, Eliza was surrounded by a hostile environment. A few people had supported her interest in science, but most had sniffed at her attempts to join the club. Science fairs had their own sense of snobbery. There was only one way to win them over: win.

Tomorrow, at judging, the battle would begin.

"Good luck, kick ass, and take names later."

At seven A.M., Katlin Hornig got a call from her dad, Bruce, who uttered his signature sendoff. By then, Katlin was up and dressed in a blue shirt, black slacks, and a blazer, with a horse pin on her lapel. Already, the eighteen-year-old from Monte Vista, Colorado, was itching to ditch this monkey suit and put on a flannel shirt, jeans, and cowboy boots. Still, looking presentable in front of a judge was important. Many science fair kids poured hours into polishing their appearance pre-judging. Katlin's only question to her mom was "Do I have any fuzzies on the back of my pants?" She was ready.

By eight-fifteen A.M., Katlin and 1,501 other competitors had marched into the convention hall. At Intel ISEF, kids weren't just up against one judge. To ensure that projects got evaluated as thoroughly as possible, competitors typically gave their presentation to anywhere from six to more than ten judges. With more than fifteen hundred students who had projects in need of judging, Intel ISEF aimed to recruit at least twelve hundred scientists, doctors, engineers, college professors, and other professionals to get the job done. It was a judge-athon of gargantuan proportions.

Although I was unable to see these epic showdowns between judges and competitors, I'd heard plenty of battle stories. Through their years of fighting for survival on the convention hall floor, science fair veterans have become adept at identifying and categorizing the different types of judges they meet. In the same way that an ornithologist can instantly spot the difference between a waxwing and a warbler, science fair kids can easily peg which species of judge they're in the presence of, and adjust their behavior accordingly. Probably the most garden-variety type of judge is the Nice Judge, who swings by with a smile, listens politely, lobs a few softball questions, then leaves, thanking students for their time. While a rookie competitor might think, *Bingo—I aced it,* any veteran knows that Nice Judges are far more dangerous than meets the eye. Politeness can signal that a judge just isn't all that excited about a project; a judge who's jazzed will pepper a competitor with questions.

While all students love a judge who's overflowing with questions, if those queries are delivered *rat-tat-tat!* and home in on a weak point in a student's defenses, that student has just come face-to-face with a different breed entirely, known as the Mean Judge. Then there's the Poker-faced Judge, who's hard to read. There's the Clueless Judge, who asks questions so out of left field, one wonders whether he or she understood the project at all. Even rarer is the Butterfly Enthusiast Judge, a subspecies typically seen flitting around the Animal Sciences category, although he occasionally ventures outside his natural habitat. One science fair competitor from Golden Valley, Minnesota, bragged that he could spot a Butterfly Enthusiast Judge hovering a mile away. "Usually they're portly, wearing plaid, have curly hair and a big, grotesque-looking beard, and they have glasses with a little string on them so they can look at your project like *this,*" he explained, pretending to hold a pair of spectacles six inches in front of his face and squinting.

Some judges were once science fair stars themselves. Robert Donato, an obstetrician from Williamsport, Pennsylvania, won second place at ISEF back in 1976 for his research on a virus that could immunize cells to cancer. Before this point, Robert hadn't given college much thought, since his parents couldn't afford it. Science fairs changed everything, winning him a full-ride scholarship to Ursinus College, which was known for its medical program. So when newspaper reporters came calling in the wake of ISEF to interview Robert about his plans for his future, he had his answer ready. "I'd like to be a doctor," he said. Robert's mother cried after reading this quote in the article the next day.

Since science fairs were a stepping-stone for everything

Robert accomplished later in life, he relishes volunteering as a judge and has been doing so for the past ten years. Many judges follow the fair to convention halls across the country, like road- ies for a Grateful Dead concert. As he trolls the aisles, Robert constantly stumbles across kids who remind him of what he was like at that age—kids who are shy, poor, and prospectless, and see science fairs as their only salvation. One year, Robert judged a boy from Japan who, based on his appearance, came from an impoverished background. Using a translator, Robert asked him questions about his project, which explored which colors attract dragonflies. When asked why he'd picked this particular topic, the boy surprised Robert by speaking directly to him in broken English. "I like dragonflies" is all he said before breaking out in a broad grin. Something about this competitor—perhaps the sim- ple sincerity of this statement—got to Robert. Later on, when the boy won a $5,000 award, Robert couldn't have been happier. But the road judges must travel to get to these happy endings was far more grueling than I would ever have expected.

By six P.M. that evening, the judging part was over. All twelve hundred judges had scrutinized their last projects, administered their last handshakes. At this point, they convened in tightly guarded back rooms to pick the winners.

Once judges were sequestered behind closed doors, they perused an interactive chart depicting the scores of projects in their cate- gory. Scores ranged from 0 to 100, but these numbers were just a starting point to help them focus on which projects might re- quire further scrutiny. From there, judges discussed why certain projects deserved to be moved up—or down—in the ranks. *The number one project has some faults I'd like to address*, one judge would say, while another would say, *I think the number four project should be moved up to the number two spot for the following reasons.* From there, other judges agreed, disagreed, or added their two cents to the argument. As projects ricocheted up and down in the ranks, the debate heated up.

"Imagine a hundred judges in a room, voting on one hundred and twenty projects. It can turn into sheer mayhem!" one judge confided in me afterward. And yet David Feinstein, a mathematician from Portland, Oregon, who'd frequented those back rooms himself, had a different take. "It's simultaneously dull as toast and sends chills up your spine," David told me. Why? "Because, my god, you're watching a group of people come to consensus on a really hard problem. You're watching adults change their mind because of reasoned discussion. Have you ever actually seen that happen in real life? Adults changing their minds in the absence of violence?" Curious how judges got the job done, David developed a computer algorithm called Group Consensus Support (GCS), which not only streamlined the judging process, but enabled him to analyze scores and track what happened during caucusing. Due to his sleuthing, he said, "You can almost begin to peer into the souls of the judges." So what did he see?

One thing David noticed was that scores alone weren't a reliable predictor of the winners. Competitors who earned the highest average score in a category ended up winning the top award less than half of the time. Occasionally a high-scoring project would receive one lone low score. But that didn't mean this low-scoring judge was off base. During caucusing, if a judge who panned a project stood up to explain why, half the time that judge would convince the plurality of other judges that he or she was right by pointing out a flaw others had missed. If the judges felt the student had made an effort to fool them and cover some-thing up, their reaction could be particularly harsh. Once, when it was discovered that one student in line for a top prize hadn't been forthcoming about the fact that his mother was his mentor, judges were so disturbed by the student's deceptiveness, he was demoted out of award range entirely.

Certain projects earned a mix of very high and very low scores, with few or none in the middle. These "polarizing projects," as David dubbed them, were often the most innovative of all. "Proj-ects that polarize judges are likely to be the real game-changers," David said. "This opinion is not only my own, but that of Nobel laureates and many others." Unfortunately, some judges gave these projects low scores, since the research headed in such pioneering directions that even they didn't know what to think. Given that two low scores sank a project's overall ranking, polarizing proj-ects ran the risk of never coming within award range. David thought that was a shame, and he developed an algorithm to help judges spot these projects so they could be pulled up from the abyss for a closer look. "It bothers me because if you drive home the lesson that innovation isn't rewarded or appreciated, they won't innovate wherever they go, whether they become sci-entists, bankers, or artists," David said. "What a tragedy for society that would be!"

If judges had a soft spot for anyone, it was the hard-knocks kid. "I do think there's a big slug of 'affirmative action' that comes in if they see a project where the quality of science is really impeccable, and the kid really rose above a background where you wouldn't expect to see that," David said. This didn't neces-sarily mean the student was in a minority group, but that he or she came from a situation that put extra hurdles in the student's path, whether that was lack of money, mentors, or access to a decent lab or library. If a student was scraping by in a garage but still managed to pull off an excellent project—which was rare, but it happened—such efforts didn't fall on deaf ears. "Judges are not immune to inspiration, and this inspires them," David said. Judges, in other words, were human. They judged with their hearts as much as their heads.

Given all the forces at play, pushing projects up and down in the ranks, which one tended to end up at the top of the heap? Typically it starts out somewhere in the top five projects and eventually emerges as the best of the bunch. The caucusing co-chair asks for a show of hands. If there's a clear majority, the decision is done. If not, judges continue caucusing. While students weren't privy to the backroom debates that would decide their fates, they often developed a decent sense of how they'd fared. Katlin Hornig left the convention hall with a hunch her therapy horses were a hit.

"During judging, you get an idea of which judges will go to bat for you," Katlin confided in me during her postmortem, adding that two judges she met seemed especially keen on her project. BB Blanchard was pretty sure she'd captivated one judge who was fascinated by the fact that she had leprosy. Taylor Wilson and his nuclear fusion reactor basked in a ton of flattering comments from judges, like "So is the government funding your research yet? This could save the country."

Not all of the students I'd been following, though, survived the judging process unscathed. During their presentation about microbe levels on infant changing tables, Sarah Niitsuma and Shwan Javdan were asked, "So what's the active ingredient in those sterile wipes that can help curb the risks?"

At this, Sarah and Shwan were stumped. As the judge drifted off, the duo racked their brains on what to say if they were asked this question again. Out of the corner of her eye, Sarah spotted the solution: a public dispenser of sterile wipes, stationed to ward off the spread of germs among judges and students. Sarah sprinted over, scrutinized the dispenser, then sprinted back with the information they needed, which soon came in handy when yet another judge asked the same question. Still other students resorted to even sneakier methods to divine the answers they needed. At one point during judging, Andy Bramante, or "Mr. B" as he was known among his Greenwich, Connecticut, students, received a text from Eliza McNitt.

What's that scripty U with a slash called?

Mr. B texted back: *Micrometers.*

Receiving his text, Eliza breathed a sigh of relief. "It's funny because for me, it wasn't the performance and delivery that made me nervous. It was the scientific jargon," she confessed to me afterward. Which was understandable: Eliza had been an actress since age four, but competing in science fairs for just the past two years. Still, overall Eliza thought she'd done well. She'd even run into a judge she'd met last year at Intel ISEF 2008 who'd been so inspired by her research on honeybees, he'd built a hive in his own backyard. Another judge sidled up to Eliza and said with a wink, "We'll be seeing you up on that stage."

Was this a tip-off that good things were in store for Eliza at the awards ceremony? Maybe she and her bees stood a chance. Maybe they'd kick some science fair butt.

As the backroom debates raged on among judges well into the night, the kids were too keyed up to sleep. Some had retired to their rooms to finish homework or study for finals. Others gathered around television screens to watch riveting (at least to them) footage of NASA repairing the space shuttle. Many more of the 1,502 competitors, though, were unwinding the usual way kids kick back and relax: at a party.

This year, the biggest post-judge party was going down in hotel room #734. The party's host, of course, was none other than Philip Streich, aka the "Next Bill Gates." Pretty much everyone had heard the story about how this home-schooled eighteen-year-old from Platteville, Wisconsin, had created a nanotechnology project that had led to five patents and a $12 million company. In keeping with his formidable reputation, Philip's hotel room wasn't just a room, but a thousand-square-foot suite complete with floor-to-ceiling windows showing panoramic vistas of Reno, Nevada, twinkling below. Philip had even driven to Best Buy and splurged on a sweet sound system, complete with subwoofer, so science fair kids could party in style.

Philip was graduating from high school this year, so this would be his last year competing at Intel ISEF. Some up-and-coming competitors, no doubt, were relieved to see him go, since he had monopolized the awards for three years. But this year, Philip wasn't all that interested in winning. His main priority was to catch up with the many friends he'd met on the science fair circuit. So every night, he and his friends plugged in their iPods, pumped up the music, and opened their doors.

As waves of kids flooded room #734, Philip couldn't help noticing one girl in particular. Just about every guy in the room turned to look, and for good reason: She was gorgeous. Many romances were sparked at science fairs, since these kids, back home, felt like "closet nerds" who had to keep their smarts under wraps. By the time they got to Intel ISEF, they were bursting with the need to meld minds, and occasionally more, with one of their own.

Philip didn't date much. Part of this was due to his busy

schedule, but that wasn't the entire story. As his pal Stephen Trusheim from Golden Valley, Minnesota, put it, when Philip first attended Intel ISEF three years ago, he was more awkward around the edges. Back then, Stephen explained, "He would have probably started a conversation with a girl, and it would have been quite lame, and the girl would have taken a 'phone call' from her 'mom' at the earliest break in words."

Time, though, could turn even the more flagrant science fair nerds into smooth operators. For one, they were adept at taking rejection without flinching. Earlier that day, they'd faced judge after judge, who'd picked their project apart. At the very least, they learned not to take criticism personally. At best, they learned how to dish out a brilliant comeback. As Stephen put it, "I'm now able to suffer through a blistering attack on my project by a judge and carefully—and factually—retort what they had to say in seconds." If that wasn't great practice for the pickup scene, what was?

Philip made a beeline for the beautiful girl who'd just walked into his suite. He introduced himself, although chances were no introduction was necessary. Everyone knew Philip. As people danced around them, she and Philip got to talking—about science, art, life. As the hours wore on and the night sky grew pale, that's when Philip brainstormed another crazy idea that, as far as this girl was concerned, was probably the smoothest line she'd heard in her life.

"Let's drive to Lake Tahoe in California and watch the sun rise," he said.

Seconds later, they were running through the hotel lobby and hailing a cab. Philip threw the cabdriver $300, which got them where they needed to go, just in time to take a seat on a hillside and watch the sun peek over the horizon. Tomorrow—or rather, today—Philip would head to the awards ceremony. Winning, after all, was what all 1,502 students had convened there to do. And yet, sitting there, with a beautiful girl he'd just met, Philip couldn't care less. Winning was the last thing on his mind.

At this point in the story, Philip, out of respect for the girl's privacy, wouldn't tell me what happened next. Which was fine as far as I was concerned. Sometimes, it's better to let the imagination run wild, rather than know all the facts. Science fairs, after all, aren't just about charts and data, or winning and losing. Science fairs are about dreams. Hopes. Friendships. Relationships. And, occasionally, the beauty of the unknown.

"AND THE WINNER IS . . ."

Winning isn't everything. But wanting to win is.
—VINCE LOMBARDI

"This is it. This is my last chance."

"There's no way I'm gonna win."

"Maybe I'll win a small award."

"Whatever happens, once I get home, I'm buying *X-Men Origins: Wolverine* and playing it for a week straight. I need a break."

On the day of the awards ceremony for the Intel International Science and Engineering Fair 2009, 1,502 students shuffled into the auditorium and took their seats. Some chattered nervously, but that was merely an attempt to distract themselves from the question front and center in everyone's mind: *Who will win?*

As I scanned the auditorium, I saw crowds of kids perched on the edge of their seats, wiping their sweaty hands on dress pants and skirts. In the back of the room, parents and teachers held their digital cameras in white-knuckled grips. The air in the room was heavy, hard to breathe. At Intel ISEF, $4 million worth of awards were up for grabs. A lucky few might rake in hundreds of thousands of dollars. Some would find out whether they could afford to go to college at all. Within the space of a few hours, kids would learn if they would one day become the doctor, astronaut, or engineer they had always dreamed they could be.

Within the sea of hopeful faces were six competitors I'd come to know well. They'd tackled seemingly insurmountable challenges to make it this far. Taylor Wilson, aka the "Radioactive Whiz Kid," battled skeptics for years before building his nuclear fusion reactor. BB Blanchard, hit with the devastating diagnosis of leprosy, fought fear with facts. Katlin Hornig convinced a bunch of burly cops to give hokey "horse therapy" a try. Sarah Niitsuma learned how to set her tiara aside, swallow her pride, and trust her science fair partner Shwan Javdan. Eliza McNitt proved she could be a blonde, beautiful, artsy actress and *still* be a science nerd to boot. Last but not least, Philip Streich, who was home-schooled on a farm in the middle of Boring Nowhere, managed to morph into a nanotechnology millionaire.

All six of these competitors were strong contenders. Still, they were only a handful out of 1,502 kids dying to win. Given those odds, none of them could take victory for granted. All they could do was take their seats, and wait.

Intel ISEF puts on an Oscar-worthy awards ceremony. Spotlights raced across the crowds. Music blasted from rock concert–sized speakers. Cameras panned the audience, projecting kids' facial expressions—nervous, excited, chatty, sleepy—onto a thirty-foot screen on stage. To kick things off, a jovial meteorologist from Reno stepped up to the podium and cracked a few jokes.

"I come here before you with one emotion primarily in my mind: absolute sheer terror," he said. "I'm not sure in the history of mankind we have ever packed so much brain power in one small room such as this. I'm a little concerned about some of the quantum physical implications. We may exceed some critical mass, and who knows, we may tear a hole in the fabric of the universe!"

Chuckles. Guffaws. Applause. More speakers approached the podium to offer their own congratulations or carpe diems, encouraging these kids to seize their role as the world's next generation of problem solvers. As proof that they had what it took, another speaker pointed out that on the day astronaut Neil Armstrong set foot on the moon in 1969, the average age of the engineers guiding that mission was twenty-six. Given that President Kennedy had issued his challenge to put a man on the moon in 1961, that means that on the day the President offered his vision, the average age of those engineers was eighteen—about the same age as the 1,502 Intel ISEF finalists sitting in this room today. As further encouragement, competitors were informed that all first- and second-place winners would get an asteroid or minor planet named after them, joining an elite club including Albert Einstein, Marie Curie, the Beatles, and other luminaries who'd had chunks of galactic real estate named in their honor.

It was enough to get any aspiring science fair star salivating. Now it was time to announce who, out of 1,502 finalists, would come out on top.

Winners for the smallest awards—fourth place—were announced first. They proceeded through the categories alphabetically, starting with Animal Sciences, then Behavioral and Social Sciences, then Biochemistry, and so on until all fourth-place winners from seventeen disciplines stood on stage. "I'm so glad I'm in Animal Sciences," one competitor confided to me, since this meant he wouldn't have to wait long to hear his verdict. "We used to be called Zoology. *That* was excruciating."

As fourth-place awardees were announced in Physics and Astronomy, Taylor Wilson heard his name called. While the rest of the kids standing on stage looked ecstatic, Taylor, I noticed, looked bummed he hadn't done better.

"I was a little disappointed," Taylor confessed to me afterward. "I think I probably had my hopes too high." After all, throngs of people, professors, judges, journalists, and even talk-show host Conan O'Brien had mobbed his nuclear fusion reactor all week. Taylor had spent two years putting it together and could recite the intricacies of nuclear physics in his sleep. All that for a mediocre fourth place? Bill Brinsmead, the technician who'd helped him build his fusor and wheel it into the convention hall, noticed Taylor's sullen mood. He offered a reality check.

"Fourth in the world isn't so bad, what are you worried about?" Bill pointed out. "And besides, there's always next year."

Bill, of course, was right. Taylor was only fourteen. This was his first year competing in science fairs. While his knowledge of nuclear physics was formidable, judges may have docked points for nuances like Taylor's lack of a detailed logbook. Such flaws would be easy to fix if he decided to take a crack at making it to Intel ISEF next year. Did he think he'd be back?

"Oh, absolutely," Taylor told me.

After the awards ceremony, Bill wheeled Taylor's nuclear fusion reactor out to the parking lot and into the back of his electric van, which was plastered with snide bumper stickers like comic book character Calvin from *Calvin and Hobbes* urinating on the word *OPEC*. It was the last glimpse this year's Intel ISEF attendees would get of Taylor's project. But I had little doubt that this wasn't the last they would see of Taylor. Next year, he would stage a comeback of Chernobyl-esque proportions. Whatever he lugs into the convention hall, it will be larger, and scarier.

Given that Taylor would be leaving Intel ISEF with $1,250 in prize money lining his pockets, I asked him if there was anything in particular he was planning to buy with it.

"Something radioactive" was Taylor's reply.

But of course.

After fourth-place winners were announced, the awards ceremony moved on to third. Then second. Then first. Students started tuning out, thinking: *I had a shot at fourth or third, but I don't stand a chance above that.* Katlin Hornig was so certain her therapy horses had struck out, she turned to the calculus homework in her lap. And yet, as first-place recipients were announced for Behavioral and Social Sciences, a German kid sitting next to Katlin gave her a nudge.

"Hey, that's you!"

In a daze, Katlin set her calculus homework aside. Seemingly in slow motion, she made her way up on stage. Out in the audience, Katlin's mom, Janet, tearfully texted her husband, Bruce, back home: *First place!* Bruce started whooping and hollering at the top of his lungs, which was quite a feat given that only half of his lungs worked. "Hell yeah! Woo hoo!" he bellowed to Katlin once they spoke on the phone.

"I was just laughing," Katlin confessed to me later. "I didn't believe it."

Katlin's first-place award came with $3,000 in cash, but that wasn't the only prize her horses pulled in. They also helped her win an $8,000 award from the United States Navy and were a crucial factor in helping her land the $160,000 Boettcher

Scholarship—a full-ride, four-year scholarship to any Colorado college of her choosing. Katlin's dream college, Colorado State University in Fort Collins, was now well within reach. Bruce could breathe easy. In spite of their mountain of medical bills and his failing health, his daughter would be fine no matter what.

While most kids in her shoes would probably kick back and celebrate after the awards ceremony, Katlin, after a modest lunch at McDonald's, was back in her hotel room doing what she did best: busting her butt. She still had to finish her calculus homework, plus she had an online Latin final that night. Then, after a few hours of shut-eye, she and her mom woke up, loaded up their gray van with their tri-panels, boxes, and duffel bags, and drove straight home, trading shifts at the wheel. Fifteen hours later, Katlin hopped out of the van and gave her dad a big hug. Then she ran out to pasture and belted out her usual call.

"Tillie-Ariel-Bubba-Belle-Jasper-Sadie-Mocha-SheDaisy-HEY!"

Within seconds, eight horses came running, heads bobbing, happy to see her. As Katlin scratched their ears, she informed them that they'd won at Intel ISEF, although she doubted they could understand what she was saying. Bruce disagreed.

"They know," Bruce said. "They can sense it."

To a certain extent, Katlin conceded her dad might be right. Given all that horses had done for her family so far, she wouldn't put it past them.

Katlin Hornig wasn't the only first-place winner profiled in these pages. Philip Streich, the "Next Bill Gates," also won first place, in the category of Materials and Bioengineering. This $3,000 prize, combined with a $3,000 award from the United States Air

Force and his other winnings over three years, brought his total earnings to nearly $250,000. While any kid would be speechless if he'd won such a sum, for Philip it was a drop in the bucket. After all, his nanotechnology research had already resulted in five patents and a company estimated to generate $12 million in sales in five years.

First place at Intel ISEF wasn't the only award Philip won that spring. He was also dubbed a Presidential Scholar. Weeks later, at the awards ceremony in Washington, D.C., Philip stepped on stage and approached the microphone with the ease of a talk-show host. He announced that someone else deserved to be up on stage with him.

"I'd like to tell you about my teacher, Ms. Amanda Streich—no, it's not a coincidence that we share the same last name!" Philip said. "Parents are often said to be one's greatest teachers and that's certainly been true for me." Amanda wasn't a scientist, or doctor, or Ivy League professor. She had no formal training in teaching at all. But while she didn't always know all the facts, she had taught Philip something far more important: to love to learn, and to delve deeply into subjects rather than coast and cut corners. Then he asked Amanda, who was sitting in the audience, to come up on stage and receive her own award.

Amanda will always remember that walk, that sensation of moving in slow motion through a sea of applause, to stand next to her son. Like many teachers, parents, and mentors, she wasn't used to being in the spotlight. Which is a shame. Maybe one day soon, that will change.

Above first place, there was an even higher honor at Intel ISEF called Best in Category. This project trumped all others in its discipline, whether that was Chemistry or Computer Science, and received $5,000 in cash. Winners would also be granted a tour of the Large Hadron Collider, the world's largest particle accelerator, located three hundred feet below Switzerland. While a subterranean tunnel might not seem like the most exciting vacation destination to some, science nerds knew better. They would kill for this trip.

Which made it all the more ironic when the ultimate anti-nerd, Eliza McNitt, heard her name called.

As Eliza strutted toward the stage, her vintage velvet dress and fire engine–red stilettos stood out from the sea of serious business suits surrounding her. But this time, Eliza didn't mind one bit. She belonged here. As a Ford model and actress, she was used to stages, and cameras, and being showered with applause. But today, she had wowed a far tougher crowd. Her research on Colony Collapse Disorder among honeybees had won rave reviews from the judges in Environmental Management. Afterward, when I asked her how it felt to be up on that stage, her response was Eliza-esque.

"It felt like winning an Oscar!" she said. Before this point, she added, "Many people didn't believe in me as a scientist. Sometimes *I* didn't even believe in myself as a scientist." But today, at this awards ceremony, she had broken through the science fair ceiling. Plus, the experience would give her a novel idea that would dramatically change the course of her college career.

After Intel ISEF, rather than funnel her $12,000 total in prize money into a massive shopping spree, Eliza invested in some professional camera equipment. That summer, she flew to Switzerland along with the other Best in Category winners to tour the Large Hadron Collider, camera in hand. This immense, seventeen-mile-long particle accelerator had recently made its Hollywood debut in the movie *Angels and Demons* starring Tom Hanks, confirming for Eliza yet again that she had misjudged science. Science wasn't dull, or dry, or boring. In the hands of an artist, science could be riveting. Maybe even her new raison d'être.

After the Best in Category winners were announced, the awards ceremony was almost over. Only one award remained, and to most finalists, it seemed so far out of reach, few dared to dream of receiving it: the Young Scientist Award. Only three students out of 1,502 finalists would win this award, and its price tag—$50,000— reflected these bragging rights. Philip Streich had won it two years earlier. But who would win it today?

Before the awards ceremony, I had asked students if they had any hunches. "It's really hard to say," said one competitor, pointing out that Intel ISEF was just too big to pinpoint the top three out of 1,502. Kids who'd won other prestigious fairs were certainly on the short list of contenders. From 2001 to 2002, Ryan Patterson, aka "Glove Boy," had made waves by becoming the first and only competitor to win not only the Young Scientist Award at Intel ISEF, but top prize at the Intel Science Talent Search competition and the Siemens Westinghouse Competition. In science fair terms, this was the equivalent of winning the Triple Crown in tennis. Ryan was unstoppable. Few competitors could have stood in his way.

This year, a seventeen-year-old from Denton, Texas, named Wen Chyan had won Siemens Westinghouse. Eric Larson, a seventeen-year-old from Eugene, Oregon, had won Intel Science Talent Search. Eric, in particular, struck many Intel ISEF attendants as a decent bet for winning a Young Scientist Award. "Eric Larson is an übergenius," kids raved. Once, to prove it, Eric and his science fair friends plunked down a particularly puzzling philosophy problem and started their timers. One kid managed to solve it in four hours, another in fifteen minutes. Eric had the answer in fifteen seconds. During Intel ISEF, I had approached Eric a few times and attempted to talk to him, but he answered in one-word sentences and drifted off, seemingly on his own plane of existence. Eric's friends told me to not take it personally. Eric was just odd. In science fair circles, his eccentricities only added to his mystique, further ramping up expectations that he was The One who'd win.

Within seconds, though, all rumors and wagers about who *might* taste victory would evaporate. Finally, after five days of nail biting and intense speculation, the real Young Scientist Award winners would be called up on stage. As the first name of three was announced, all 1,502 competitors held their breath.

"From Charlottesville, Virginia: Tara Anjali Adiseshan!"

Trumpets blared. Waves of applause carried Tara forward, followed by the next two winners: Olivia Schwob from Boston and Li Boynton from Bellaire, Texas. Up on stage, surrounded by camera bulbs popping, confetti falling, all three stood there, clutching their plaques. To my surprise, they didn't look triumphant. They looked small, young, shell-shocked. At that moment, it hit me that in spite of the $50,000 they'd just won and the fact that they'd just been dubbed our nation's next wave of Einsteins, they were just kids. When not solving the world's problems, they played Guitar Hero or hung out at the mall. Parents of Young Scientist Award winners are often more stunned than their kids. In the audience, cameras drop to the floor. Video footage shakes uncontrollably, while the audio fills with the sound of parents sobbing *Oh my god* over waves of applause.

I'll admit it: Part of me was disappointed, albeit not surprised, that none of the kids I'd been following were up on that stage. Still, during my interviews with these three top winners, I noticed that their stories struck many of the same chords I'd heard already. Fourteen-year-old Tara Adiseshan, whose project explored the evolutionary relationship between sweat bees and microscopic organisms called nematodes, was home-schooled by her mother. Like Philip Streich, she was proof that kids didn't need to attend a top school to succeed. Sixteen-year-old Olivia Schwob had genetically engineered smarter worms by implanting a mouse gene called GAP-43. But Olivia didn't consider herself a genius, just a hard worker, echoing a sentiment shared by most kids I'd met.

Only how does winning $50,000 and being dubbed one of the world's most promising young scientists *feel*? This question, more than any other, left all three of them tongue-tied. "It's still sinking in," they told me. But seventeen-year-old Li Boynton probably put it best.

Li was endearingly up front about the fact that she was insecure. She seemed to shrink away from journalists as she spoke. Her wispy voice conveyed that she was uncertain whether anything she said was right. At a science fair a year earlier, one judge had done her a favor by pointing this out. "Your project was very good," the judge said, "but your presentation was not fluid. You seemed so nervous that I doubted whether or not you really knew what you were talking about." To improve her oratory skills, Li had joined her high school debate team. But her self-esteem remained tougher to fix.

During her junior year, Li started a science fair project using bioluminescent bacteria to test for contaminants in water. Armed with stronger debate skills, she clambered up the science fair pyramid from her school's local fair to regionals, then regionals to state, qualifying for Intel ISEF 2009. Just making it here was a thrill for Li, who walked into the awards ceremony fully expecting that she wouldn't win anything. Which is why she was shocked, perhaps more than most, to hear her name called as the third winner of the Young Scientist Award.

As Li stood up on stage in front of thunderous applause, she felt the grip of her insecurities loosen ever so slightly. And that feeling stuck, even after she flew home. Days later, on a whim, Li Googled her name. Normally, very few links would have popped up. Now there were pages and pages of news articles about her accomplishments. But far more important than this one triumph were all the little victories she'd worked for and won along the way. In the lab, she had learned to forge on through countless rounds of trial and error. To prepare for judging, she had learned how to present herself with confidence, which wasn't an inborn trait as she'd once thought, but a technique that improved with practice. Li had also learned, by discussing her research with judges, that her work was worthy of merit. Science fairs had opened Li's eyes to the fact that she had something to contribute to the world.

"I often thought of myself as an imposition, like my existence was detrimental to others," Li admitted. "But now after science fair, I think of myself as significant, as if I can actually do something that will help people." The experience also gave her the courage to take risks she wouldn't have dreamed of taking otherwise. "I'm applying for college this fall," she explained. "Before I was shooting for Northwestern, maybe Brown, although my dream schools have always been Yale, Columbia, and Stanford. Now I'm definitely going to apply for those schools." A year later, Li would be accepted to Yale and decide to attend.

Science fairs open kids' eyes to what they can do. The fact that all three Young Scientist Award winners were girls in 2009 might strike some as surprising, but not to the judges or staff who run these events. Good science is good science. It doesn't matter if you're male or female, black or white, wealthy or destitute. Winning changes kids. And not all of these changes can be measured in a check, plaque, ribbon, or even whether the kids go to college. Some of the most significant changes are far more subtle. Winning opens their eyes to a world of possibilities. It nudges them to take risks. It turns on that little voice inside their heads that says, *You can do this,* even when they swear they can't. It gives them grit, and guts, and the knowledge that they have the smarts and heart to handle whatever life throws their way. It gives them hope that, in spite of the odds, they have what it takes to end up on top.

BB Blanchard's project on leprosy did not win an award at Intel ISEF 2009. And yet unlike many science fair competitors, who headed home disappointed or even in tears, BB and her partner, Caroline Hebert, were as bubbly as ever.

"Caroline and I were not really expecting to get an award," BB told me afterward. "We went for the experience, and we both agreed that it was one of the best experiences we have ever had!" Winning, after all, had never been BB's main goal. She wanted to show the world that having leprosy was nothing to be ashamed of. Just by showing up, she had aced the job.

After Intel ISEF, BB continued to speak out about leprosy, in unlikely places. During a church mission trip to a nearby town, she and her youth group painted houses and built handicap ramps. At the end of the day, kids sat in a circle and talked about their faith and how it helped them overcome challenges. BB talked about how two years ago, she'd been diagnosed with leprosy.

She confessed that for a while, she was horrified. But now she understood why. Since then, she had helped remove the stigma surrounding her disease. She had turned it into an award-winning science fair project and an amazing trip to Intel ISEF. It was the ultimate lemons-into-lemonade story, and BB had her upbeat outlook to thank. "Some people live with regret or become upset and don't do anything about their problems," BB told me. "I'm the type of person that has always lived through problems and gone on with my life, trying to find the positive aspect of every setback I've faced."

Another blessing in BB's predicament, of course, was science. Researchers at Carville, the first leper colony in the United States, had not only found a cure for leprosy, but fought hard to overturn its reputation as a biblical scourge. Had BB been born in a different time and place, her fate might have been much more grim. Her mother, Anne, was reminded of this fact one day while cleaning out the school library before summer.

"I came across a book called *Dark Light*," Anne said. She opened it. It was a historical novel about a girl named Tora who lived in Norway in the nineteenth century. At the age of thirteen, Tora was diagnosed with leprosy. As a result, she was taken from her home, away from her family, and imprisoned in a leper colony until her death.

Anne couldn't help noticing that Tora was about the same age as BB when she'd been diagnosed with leprosy. Thankfully, that was just about the only thing Tora and BB had in common. So much could change in two hundred years. With BB, so much had changed in two years.

Sarah Niitsuma and Shwan Javdan did not win an award at Intel ISEF 2009. But Sarah's dream of winning a scholarship and attending college did come to pass. Thanks to their Grand Prize at the regional Salt Lake Valley Science and Engineering Fair, Sarah and Shwan won an $80,000 full-ride, four-year scholarship to Westminster College.

For Sarah, winning this scholarship meant everything. But looking back, she thinks that's just one of many things she learned from science fairs, a point she made poignantly a few weeks later during her high school graduation speech. Sarah wasn't the valedictorian or salutatorian; the Academy for Math, Engineering & Science allowed all students to give a speech if they wished, regardless of class rank. As Sarah stepped up to the podium, she was wearing a tiara perched in front of her graduation cap. It soon became clear why.

"When I first started AMES, I considered myself as a princess," Sarah said as her opener. She thought she could glide through school without trying too hard or getting her hands dirty. Science fairs were her reality check. In Tanya Vickers's science class, she learned that by working hard, she could rise above her past, and that even when the pressure became unbearable, she had the strength to keep going. Science fairs also taught her how to let down her guard and trust people, like her science fair partner Shwan Javdan. "That was a huge step in my learning process: learning how to work with somebody," Sarah admitted. In sum, she continued, "I can tell you that I am not just a princess anymore. I have been challenged. I have learned through my experiences. And I have traded my crown in for a lab coat."

At this point, Sarah took off her crown and put on a lab coat. The audience cheered, and cried, but few cried harder than Sarah's science teacher Tanya Vickers and the high school principal, Dr. Al Church. After all, they'd been keeping a careful eye on Sarah ever since she first floated into their school, topped in her tiara, and shared with them her dream of going to college. Thanks to science fairs, that wish would become a reality.

"We always cry, don't we?" Dr. Church joked with Tanya. "We're the most effusive criers on staff."

It was true. Sometimes it was just too much, watching these kids grow up.

CONCLUSION

Don't cry because it's over. Smile because it happened.

—DR. SEUSS

The instant the awards ceremony for the Intel International Science and Engineering Fair 2009 was over, 1,502 competitors poured out of the auditorium and drifted back to their booths. It was time for their final task: to pack up their tri-panels and head home. Only unlike during setup, when the tension was thick and excitement levels high, now students were clear-eyed and calm, as if they'd survived a storm that had passed.

Science fairs train kids to see the upside to any situation. A perfect example of this was Nicholas Christensen, a seventeen-year-old from Wetumpka, Alabama, who was born severely hearing impaired. Nick's mother fretted that her son's disability would hold him back. Even with a hearing aid, face-to-face conversations were a struggle for him. On the school bus, jokes flew over his head. Phone conversations were especially difficult for him to decipher, so when girls started calling, Nick rarely called them back, sticking mostly to email and instant messaging. And yet in spite of these hurdles, Nick liked to think that being hearing impaired gave him an edge. For instance, by turning off his hearing aid, he could tune out unnecessary distractions—annoying teachers, boring classmates—and concentrate, which was a boon for getting stuff done. His idol, Thomas Edison, was also hearing impaired. That hadn't stopped him from discovering many things. Maybe it had helped.

Nick had never liked wearing his hearing aid, where sounds were easily distorted. So when science fair rolled around during high school, he set his mind to inventing a better model. During Intel ISEF 2009, Nick unveiled his new, improved hearing aid, EarMeNow, which slowed down sound waves rather than amplifying them, improving word recognition in hard-of-hearing subjects by an average of 11 percent and up to 60 percent. Nick won second place and more than $50,000 in cash and scholarships. He plans to apply for a patent. Had Nick not loved participating in science fairs—or for that matter, had he not been hearing impaired himself—he never would have experienced the successes he was relishing today.

Nick was typically a shy, straitlaced kid. But science fairs also helped him come out of his shell. During the awards ceremony, up on stage, Nick had stuck out his tongue at the audience. His mother was horrified, until her son set her straight.

"Mom," he said, rolling his eyes, "it was my tribute to Einstein. Remember that classic picture of him?"

Oh. Of course. Any science fair kid would have known that.

Days, then weeks passed. Science fair season faded into summer, then fall. What were the six competitors I'd followed during Intel ISEF 2009 up to now?

. . .

The first thing I noticed upon calling Taylor Wilson, the "Radioactive Whiz Kid," was that his Southern drawl had dropped a few octaves. Taylor, now fifteen, was growing up, and busier than ever. Over the summer, his older sister, an event planner in Washington, D.C., had pulled some strings to get her brother meetings with high-ranking officials at the Department of Homeland Security and the Department of Energy (one of whom was a former science fair winner herself). The fact that Taylor was only fifteen didn't faze them. In fact, they were thrilled to see someone so young take an interest in nuclear science.

"Most of the Cold War nuclear scientists are dwindling," Taylor explained. "My closest competition is fifty years old. Nobody my age is interested in this. So it really comes as a shock to these people. I think they were a little lenient when they heard they would be meeting with me, but once they got in there, they were really excited that, even though I'm fifteen, I really have some good ideas."

Sitting with Taylor around conference tables, government officials discussed how his research could be used to detect radioactive materials that terrorists were trying to smuggle into the country through cargo containers at ports. Current methods, they said, could stand improvement. If Taylor put together a grant proposal, they'd be happy to consider funding his work. Taylor? Fight terrorists? He left Washington, D.C., elated. After two years of putting up with people's doubts, he'd finally convinced someone—and not just anyone, but high-ranking government officials—that he was part of the solution rather than a problem.

In May 2010, Taylor flew to San Jose, California, having qualified yet again as a finalist for the Intel International Science and

Engineering Fair. This year, he improved on his fourth-place win a year earlier by winning third place in the Physics and Astronomy division, as well as awards from the United States Army, GE Energy, and an all-expenses-paid trip to tour the Large Hadron Collider in Switzerland. His parents, Kenneth and Tiffany, couldn't have been prouder. To this day, they still worry about Taylor—they're parents, and parents worry—but their son has won their trust. As proof, I spotted a photo on Taylor's website (Sciradio active.com). It was of Taylor, wearing a gas mask and rubber gloves, hovering over a table of bubbling flasks belching plumes of smoke. Taylor was turning uranium ore into yellowcake, a precursor to the uranium metal that's used in nuclear weapons and power plants. The gaseous cloud surrounding him contained hydrofluoric acid, which he'd extracted from a rust remover called Whink. Hydrofluoric acid, if inhaled, can cause cardiac arrest. If a speck landed on Taylor's skin, it could corrode through his tissue, enter his bloodstream, and kill him. Thus the gas mask and rubber gloves.

While most kids might get grounded for months if caught conducting such an experiment, Taylor's dad, Kenneth, had ventured nearby to snap a few photos to commemorate the occasion, much like other dads might take pictures at Little League. No, it's not what Kenneth ever imagined he'd be doing as a dad. But he wouldn't want it any other way.

Down in Baton Rouge, Louisiana, BB Blanchard had set her mind toward an equally ambitious project that summer: finding a boyfriend. Since her diagnosis of leprosy, no one had been beating down her door for a date, but BB didn't back down easily. Plus she had her eye on the perfect guy. His name was Gunter.

BB and Gunter had been friends since they were kids, hanging out at pool parties at the Village St. George Aquatic Club. Gunter was the lead guitarist in a band called the Love Handles and an avid Ultimate Frisbee player. He'd even taught BB how to throw a Frisbee—which, for BB, was no small feat. While they often joked that they would eventually marry each other, Gunter assumed BB would never consider him boyfriend material.

But by the end of her sophomore year, BB found herself warming up to Gunter. In an odd sort of bonding moment, Gunter confided that he, like BB, had a rare disease. His was called familial mediterranean fever, which caused joint inflammation. BB swooned. "We're perfect for each other," she gushed to her friend Catherine. Only so far, Gunter hadn't made a move.

One day, while BB and Gunter were waiting in line to dive down a Slip 'n Slide, BB tossed out, "So what do you think about us?" before she dove, leaving him to mull her proposal over. Weeks later, he showed up at St. George with a duffel bag. Spotting it, BB was puzzled. Typically guys didn't bring much more than a towel to the pool. Then Gunter walked up to BB, opened the bag, and pulled out a rose.

"Will you be my girlfriend, finally?" Gunter asked.

Elated, BB said yes. They've been dating ever since.

In the fall of 2009, Bruce Hornig uttered his signature sendoff one more time—"Good luck, kick ass, and take names later"—as his daughter Katlin left home for her freshman year at Colorado State University in Fort Collins. She is currently majoring in animal sciences. When I called her, I heard what sounded like a hundred sheep bleating in the background. "They're giving birth," Katlin explained, adding that she was on call to help them deliver. "Can I call you back later?"

Someday, Katlin hopes to continue her science fair research

on the healing power of horses. Her timing couldn't be better. Horse therapy, also called equine therapy, is growing in popularity. American soldiers traumatized during their posts in Iraq can now find comfort in the Horses for Heroes program in twenty-two states. In hospitals, miniature horses known as emotional support animals (ESAs) make the rounds, lifting patients' spirits. Kids with cerebral palsy, autism, and other disabilities flock to horse therapy programs across the country. Katlin seems to be suffering from a bit of horse withdrawal herself. After becoming friends with her on Facebook, I spotted a status update from her that summed up her malaise: *Katlin Hornig misses her horses.*

Back home, whenever the cops hit a rough patch, they know exactly who they can call. Bruce is happy to drive Katlin's therapy horses to police stations, rehabilitation centers, foster homes, senior living facilities, and wherever else they can do some good. He's seen old ladies plant big wet kisses on the horses' velvety noses and grown men sob into their manes. Bruce never needed science to tell him that horses were good for his health. He was a living, breathing example of this himself. And he hopes to remain one for a very long time.

Sarah Niitsuma began her freshman year at Westminster College on full scholarship. After nearly a decade of crashing on other people's couches and sitting on toilet seats to finish her homework, she got her own apartment with a friend, which thrilled her to no end. She did well in her classes, especially math and science, although she sometimes still struggled to put her past behind her. On December 19, 2009, she was arrested for distributing ecstasy at a rave. Sarah, ever the optimist, said she's actually glad she got caught early on during her college career, since it scared her into shaping up her act.

"I met with the dean of my college and told him everything that had been going on, and that I was sorry, and that I wanted to change," Sarah told me. "He ended up giving me another chance." So far, Sarah has lived up to her promise. "I've made a lot of decisions that aren't great," she admitted. "But I try hard. I'm trying to get everything back on track."

Shwan Javdan, now a high school junior, started working on another team science fair project under Tanya Vickers' tutelage. Based on her experiences as a science fair coach, Tanya published the *Teen Science Fair Sourcebook,* hoping to show more kids what it takes to do well, and win.

Sarah and Shwan are still friends. They don't talk much, but Sarah knows that if she ever needed anything, Shwan wouldn't hesitate to help her out. And she'd be happy to return the favor.

Eliza McNitt started her freshman year at New York University's Tisch School of the Arts. While she had planned on studying acting, science fairs gave her a change of heart. Rather than star in films, she decided to create films herself.

Her first project was *Smashing Science,* a documentary about the Large Hadron Collider in Switzerland, which she had toured that summer as part of her prize at Intel ISEF. "Intel was my producer, in a way," Eliza joked. But that's hardly the only impact science fairs had on her life. "If it weren't for ISEF, I wouldn't be studying science in school," she said. Eliza is double majoring in environmental science and documentary filmmaking.

Eliza ended up a science fair star by accident. But she hopes her story, and her films, will convince more kids to set aside their preconceived notions about science and give it a try. "One of the things I definitely learned is that science fairs are not just for nerds," she said. "I definitely I walked into it with a bit of bias and walked out with many excellent friends." She also learned that not fitting in—with scientists, or artists, or any particular group— might be lonely at first. But ultimately it can give you an edge. "It definitely distinguishes me in the art world and in the science world," she says.

There's even hope for the honeybees. Currently a vaccine called Remebee, fed to bees through sugar water, has shown promise in curbing the effects of Colony Collapse Disorder and bringing bee populations back to life. Eliza hopes to continue her publicity efforts on the bees' behalf. Every day, she wears a gold bee pendant around her neck as a reminder of their importance in our lives, and in her life in particular. Without bees, Eliza's college career might have veered down an entirely different path. "One of my friends joked that bees are the gift that keep on giving," Eliza laughed. People underestimate bees, just as they once underestimated Eliza's abilities. But finally, for both Eliza and the bees, the future seems bright.

Philip Streich started his freshman year at Harvard University. Even here, his reputation preceded him. At parties, strangers would say, "I know who you are. You're Philip Streich, the next Bill Gates!" Philip did his best to downplay these rumors, but it was no use. Science fairs had made him famous.

In between his classes, Philip found time to open up a satellite branch of his company, Graphene Solutions, where he's currently working on developing an inexpensive graphene-based solar panel. He also founded a record label, 91 Records, which now represents artists across the country. Rather than sticking to the usual Harvard haunts like Finals Club parties, Philip prefers to throw his own parties, which are always mobbed with kids.

At Harvard, Philip met many graduates from Exeter and Andover—schools he once pined to attend. Philip smiles to think how different his life would have been had he gone to one of those schools. "I'm sure I would have loved it, but I don't think my life would have been the same adventure," he explained. Out on a farm in Wisconsin, in the middle of Boring Nowhere, he had learned everything he needed to know. He had learned things he couldn't have learned anywhere else. Philip will never forget his first day of high school, at home when his mother asked, "So what would *you* like to learn today?" No teacher had ever asked him that. Only why not? Maybe, to succeed, that's all kids need.

ACKNOWLEDGMENTS

While writing this book, I met many science fair competitors whose accomplishments are not chronicled in these pages. Nonetheless, I was amazed by their stories and grateful for the time they took to open their homes, laboratories, and lives to me and answer my many questions. I owe a special thank-you to Gabriel Joachim and his parents, Lorna and Franz, whose passion for science fairs infuses every page of this book. Gabe didn't win an award at Intel ISEF 2009 but would end up winning Best of Category at Intel ISEF 2010, proving to me that hard work eventually pays off, provided you don't lose heart.

I was also inspired by Fatimah Alkhunaizi, a Saudi Arabian girl who dreams of skydiving, and her mentors Saeed Saeed and Amr AlMadani, whose after-school science program, the National Talent Training Center, is encouraging girls and boys across Saudi Arabia to dream big and take risks. I am thankful to Chad Holloway, who battled bumper-to-bumper traffic with a sick daughter to tell me in person how science fairs changed his life.

I am amazed that Siyabulela Xuza managed to find me amid Penn Station madness to tell me his story. And Stephen Trusheim and Sujay Tyle went above and beyond, showering me with leads, ideas, and funny stories about the science fair world.

I am also grateful for the time and insights of the following students: Tara Adiseshan and her mother, Mamtha, Apas Aggarwal, Breanne Anderson, Jorge Aviles, Scott Betz, Nikita Bogdanov, Ian Boulighy, Joe Bussenger, Kieron Callahan, Amy David, Jose Davis, Erika DeBenedictis, Omar Figueroa, Eric Foss and his father, Gary, Dylan Freedman, Karen Garcia, Michael Gord, Haijun Guo, Amber Harris, Marley Iredale, Taylor Jones, Kristen Kearney and her mother, Kathy, Alexander Kendrick, Muhammad Naeem Khan, Martin Lopez, Suraj Mishra, Felix and Natalie Nguyen, Olivia Schwob, Aaron Scruggs, Divya Shenoy, Anna Simpson, Janet and Benjamin Song, Nilesh Triperani, Sheel Tyle, Mario Valdez, Pablo Valdivia, Marlene Weikle, Caroline Wurden, Quing Xu, Ao Yang, and many others.

I am also indebted to the Society for Science & the Public (SSP), the organization that runs the Intel International Science and Engineering Fair (ISEF) and Intel Science Talent Search (STS). Thank you, Rick Bates, Marisa Gaggi, Michele Glidden, Nancy Moulding, and Sharon Snyder, for placing your trust in me. I hope this book conveys my admiration for your work. Phillip Huebner, Robert Yost, and Bonnie Blazer-Yost were also amazingly helpful putting me in touch with students to profile in this book.

A writer is nothing without a talented editor, and I feel incredibly lucky to have had Sarah Landis from Hyperion at the helm, deftly offering solutions to problems that had me stumped. Likewise, my agent Douglas Stewart at Sterling Lord Literistic is my own personal miracle worker. His instincts and advice proved invaluable from the conception of this book to its finish.

ACKNOWLEDGMENTS

Kiri Blakeley, Carol Huang, and Elise Nersesian provided critical editorial advice and plenty of morale boosts as I wrote this book, teaching me just how rare and wonderful good friends are in this world. And no one deserves my gratitude more than Jason Kersten. Thank you for all the miles you drove on this book's behalf, for reading my draft, for telling the truth, and for teaching me when it's time to shut the laptop and enjoy life a little. It's a lesson I'm still learning, but it's the most important lesson of all.